Advanced Conversational Hypnosis Selected Works

A Collection of selected works focusing on advanced conversational, Ericksonian hypnosis techniques and skills taken from 'Learn Hypnosis & Hypnotherapy From Beginner to Mastery: Becoming A Brief Therapist Special Edition The Complete Works Vol I-IV'

By

Ericksonian Hypnotherapist

Dan Jones

Contact the author:

www.facebook.com/wellbeingcoachdanjones

First Edition 2011

Published By Lulu.com

Copyright © Daniel Jones 2011

Daniel Jones asserts the moral right to be identified as the author of this work

All rights reserved. No part of this publication may be reproduced, stored in a retrieval system, or transmitted, in any form or by any means, electronic, mechanical, photocopying, recording, or otherwise, without the prior written permission of the publishers or author.

ISBN: 978-1-4709-2429-4

1FIRST EDITION1

Contents

Advanced Conversational Ericksonian Hypnosis Skills: Selected Works .1

How to Induce a Trance 7

What is Hypnosis? 11

How Do You Do Hypnosis? 17

Post Hypnotic Suggestions 29

Hypnotic Language Patterns 33

Illustrations of the Use of Hypnotic Language Patterns 51

Gestures and Internal Reality 55

Erickson's Early Learning Set 59

Hypnosis & Trance 61

Working with Ideo-Dynamics 65

Subliminal Auditory Stimulation 69

Learning to Notice Minimal Cues 73

Hypnotic Language Patterns, Skills and Ideas for Working with People 77

Breathing and Minimal Cues for Deep Rapport Building 87

How to Match and Mirror Successfully 91

My Friend John Technique 95

Putting Yourself in an Externally Focused Trance 97

Peripheral Vision 99

Not Doing to Create Change 101

The 'Just Being There' Trance Induction 103

Relaxation; Trance and Trance Signs 105

Creating Dissociation, Metaphors and Age Regression 109

Using the Crystal Ball Technique 111

Hypnosis, Trance Induction & Utilisation 117

My Experience of Stopping Using Scripts 121

Describing Your Own Experience to Induce a Trance 125

Arm Levitation and Catalepsy 129

Hallucinations 133

Surprise and Confusion 137

Compound Suggestions 139

Contingent Suggestions 143

When are People in a Trance Naturally? 145

Binds, Double Binds & Implication 147

Post Hypnotic Suggestions (PHS) 149

Nominalisations 151

Fractionation 153

Inducing Trance with Music 155

Every day Trance Phenomena 157

Rapid Inductions 159

Observation Skills 161

Analogue Marking 165

Rapport 169

Therapy in Action – Performance Enhancement: A Conversational Hypnosis Demonstration 171

   Initiating Trance 173

   Seeding/Priming Using Metaphor 175

   Induction 182

   Trance Deepener 187

   Contingent Suggestions, Compound Suggestions & Nominalisations 189

   Open Ended Suggestions 195

   Metaphors for Unconscious/Conscious Processes 197

   Post Hypnotic Suggestion 198

   Shutting Down Perceptual Filters 203

How to Induce a Trance

Hypnosis can be induced by focusing your attention (could be on a spot on the wall, or on a thought, or on a rhythm, or on almost anything else) which is what happens when the reorientation response is fired.

Hypnosis or trance states can be induced in many different ways:

- By confusion followed by a solid suggestion.

- Pattern interruption, (like handshake inductions) these fire the reorientation response as the correct pattern isn't happening so they take their cue on what to do next from the hypnotist.

- Shock inductions (like most stage hypnotists do, these set off the reorientation response).

- Relaxing the muscles (which are part of the process for falling asleep).

- Deepening rhythmic breathing (part of the process for sleep).

- Visualisation (part of falling asleep and dreaming).

Everyone uses hypnosis all of the time. People think about winning the lottery and what they would do with the money, they are visualising which induces a light trance. Smokers may go into a trance when they focus on the cigarette they are having and they take deep breaths as they allow their muscles to relax. When people have cravings they enter a trance as they as so intensely focused on what they crave. When people get angry they focus on what is causing the anger. When people get depressed they focus on worrying and negative thoughts.

Doctors, counsellors and other psychological therapists all use hypnosis all of the time, often without realising it. It is when they don't realise that they can cause more harm than good.

For example: when a doctor is seeing a patient, that patient is in a mild trance state, usually a slight anxiety trance. Their whole focus is on the doctor and on what the doctor has to say. If the doctor gives any suggestions they will be acting like a Hypnotherapist so it is important that they give good suggestions. If they say 'this is going to hurt' it increases the chances of causing pain. If they say '80% of people die from the cancer' they are suggesting the patient is unlikely to live.

Many counsellors and psychotherapists that don't realise they do hypnosis can give equally harmful suggestions that make clients leave sessions feeling awful. Whenever a client leaves a session they should feel empowered, they should feel like they have achieved something in the session. They shouldn't leave feeling worse than they came in. It is the therapist that has the power to make the client feel better or worse.

Each time a doctor, counsellor or other psychological therapist asks a patient to think about something they are making the patient visualise which is causing them to enter a light trance. Depending on what they are being asked to think about each thing they think of will have the effect of updating current patterns in the brain. This is why it is important to have patients think of things that desirably adjust patterns rather than getting people to keep thinking about their problems.

Hypnosis CAN be induced in anyone

It used to be thought that not everyone could be hypnotised but this was because in the past hypnotists would use a script that was the same for each person. This didn't work on everyone because people are all different. For example, some people might feel uncomfortable with an induction that guides them down in an elevator so they won't respond by going into a trance. Now well-trained Hypnotherapists will tailor the induction to the specific client and let clients go into trance in their own way.

What is Hypnosis?

Hypnosis is a trance state. Trance states involve a narrowing focus of attention.

This could be:

- Outwards like in an emergency.

Or

- Inwards like when daydreaming or worrying.

A hypnotic trance state is when you access the Rapid Eye Movement (R.E.M) State. This state is accessed during dreaming and at times when the brain doesn't know what is coming next, like in an emergency or with a loud noise.

The R.E.M state is the state that you go into to create or update patterns of behaviour. This is why human babies have the highest time in an R.E.M state in the three months leading up to the birth. In this last three months all of the instinctive patterns are being laid in place for life on the outside. This allows for certain behaviours to happen without being learnt, like breathing, suckling and the ability to match facial expressions which allows the baby to bond by building rapport.

There are a number of behaviours associated with trance states many of which are useful to be used for rapid healing.

Trance state behaviours include:

- An increase in suggestibility & responsiveness
- Increased tolerance to pain
- Hallucinations
- Immobility
- Blinking stops (sometimes it starts rapidly first)
- Ability to change body temperature
- Ability to build muscle using the imagination
- Ability to alter blood pressure
- Ability to change mood
- Ability to rehearse new behaviours & make them instinctive
- Altering immune system activity

- Accelerated healing
- Amnesia

Before learning how to induce a trance in yourself and others it is important to know what to look out for. If you don't know what to look for to tell when someone is in a trance you wouldn't know when they are hypnotised. The ability to help people into an optimum learning state, which is the same state as a hypnotic trance is one of the most important abilities that you can learn.

When you know what to look out for you can begin to utilise what you see as being an indicator that someone is entering (or is in) a trance. You can also then begin to observe people in everyday life and notice natural trances people are in.

Trance indicators include

- Catalepsy
- Different voice quality
- Shorter sentences and words
- Relaxed muscles
- Less body movement
- Economy of body movement

- Smoother features
- Lack of startle reflex
- Takes things literally
- Slow or no swallowing reflex
- Slow or no blinking
- Slower pulse
- Slower respiration
- Pupils change
- Head nodding side to side
- Facial symmetry
- Breathing from stomach
- Less facial colour
- Eyes roll back
- Eyes flutter
- Instant hypnotic phenomena

Not all of these indicators happen all of the time. Sometimes some people may show some indicators but not others or there may be a delay before some responses. This delay can often happen with hypnotic phenomena or tasks that clients are asked to carry out. This happens because often internal time distortion occurs sometimes on an unconscious level that can make the time it takes for a client to carry out a behaviour seem quicker to the client than it appears to the therapist. There can also be a delay due to needing some processing time to 'action' what they have been asked to do.

How Do You Do Hypnosis?

To do hypnotic induction's you need to either recreate stages leading to dreaming sleep or recreate the state of not knowing what is happening next causing the reorientation response.

Recreating stages of sleep could be a relaxation induction getting the client to relax their body perhaps starting with their feet, and then relaxing their mind by getting them to think of something pleasant. Or it could be getting them to imagine something relaxing. Or getting more of their attention focused inwardly in some other way.

Recreating a state of not knowing what is happening next could be done by interrupting a pattern of behaviour, or causing confusion.

Some types of induction are:

- Conversational (overt & covert)
- Pattern interrupt
- Embedded-meaning/metaphorical
- Confusion
- Directive

Conversational inductions are induction's that initially start with an ordinary conversation. They involve embedding suggestions and utilising on-going experiences or events to induce a trance.

It could be embedding suggestions in a conversation or feeding back what a client says to deepen their experience.

An example of a **conversational induction**:

As you sit back and **begin to feel comfortably relaxed** (Embedded command), I would like you to **let those eyes gently close**...that's right...recognising that with those eyes closed you can **go inside very pleasantly**, accessing memories, past experiences or other meaningful events, times gone by when you felt good... Now, Graham, I'd like you to take two deep, refreshing breaths and as you release that second breath you can **drift even more deeply** into a satisfying a pleasant state of relaxation...etc.

An example of a **pattern interrupt induction**:

(Interrupting the pattern of a handshake)

Hi, I'm Dan (hand goes out; clients hand comes to meet it. I take it with my opposite hand, raise it with palm facing clients face then slowly start it moving to their face)...and as that hand continues to move closer to your face all by itself you can begin to notice the change in your vision...and as the vision changes you can notice how heavy those eyelids are getting...and you won't go all the way into a trance until that hand comfortably touches the face...etc.

An example of a **metaphorical induction or embedded-meaning induction** would be to tell a story and use embedded commands and metaphors for going in to trance...etc.

An example used in a staff meeting to get the staff working together again:

One-day snow white decided that she wanted to go on a walk, she didn't often go out far from her home as she was unsure what she would find in the deep, dark forest. Snow white left on a path right outside her front door. The path was covered by trees arching high overhead; either side of her was deep, dark forest. Snow white stuck to the path walking through the shimmering beams of light that flickered down through the trees above. As she continued to...*follow this path*...she was aware of the rhythmic beat of her feet on the ground and the sounds of birds in the trees and the rustling of leaves as the wind blew a breeze. She continued to wander and at times found her mind wonder about why she set out on this journey...after walking for a while she found herself smile as she saw a house in the distance. The house was in a clearing in the forest that was bright and cheerful. There were plants of many varieties and many flowers surrounding the house. As snow white reached the clearing she could *feel the calm*, warmth from the sun on her skin. Snow white could hear voices coming from the house and the closer she got the more she could tell that the people inside the house were disagreeing with each other. Snow white approached and asked one of the people what was wrong. Grumpy explained that they used to all go to work singing and dancing with enjoyment but now they seem to have forgotten how to work as a team. Grumpy explained that they used to push together...**pull together**...axe together...**all together**...but now they found that they couldn't. When one pushed another pulled and no work got done. Snow white asked what they do and was told that they are the team that digs and lays the foundations for new buildings. She asked them why they decided to do that work. She was told that you see buildings standing and feel proud because you know that they are standing because you built the foundations well, it makes you proud of all that hard work you did...snow white decided to tell the little people a story about a centipede that kept falling over its legs. The centipede asked a friend how he manages to walk without falling over. He was told to just...**relax**...and let all the legs...**work together**...not keep thinking about which leg should do what and when. This made no sense to the dwarves so they decided to forget what snow white said and just enjoy her company. Before snow white left she asked who made such a lovely garden. The dwarves said they all worked at it and that many of the plants have survived some harsh winters. At the end of the day snow white said good bye to the dwarves. She got right up and left. As she left she was amazed by how much happier and healthier they were starting to become. Something had happened that they were learning from which looked like it made them healthier and made them work out their differences, sneezy had stopped sneezing, grumpy was happy, bashful had clear skin and no hint of red, and all of the others had noticed improvements too. This made snow white happy as she skipped away from the house up the path leaving her adventure behind like a dream that got more out of reach like a name on the tip of your tongue as she approached her home pleased with her mini adventure, then walked through her gate and, finding it was all a dream she...***opened her eyes***...

A directive induction is an induction where you tell the client what to do.

An example of a **directive induction**:

I'm going to shake your hand three times...the first time your eyes will get tired...let them...the second time they'll want to close...let them...the third time they'll lock and you won't be able to open them...want that to happen, and watch it happen...now...1...2...now close your eyes...now 3...and they're locked and you'll find they just don't work, no matter how hard you try...the harder you try the less they'll work...test them and you'll find they won't work at all...

An example of a **confusion induction** (used within a story):

One afternoon a woman set out looking for her friend's house. She was feeling rather tired and sleepy, but perked up halfway there when she realised she'd forgotten the directions. She decided to check for directions anyway, and holding the wheel with her right hand she used her left hand to place a can of coke on the floor right beside her then reaching right across her side with her left hand to her right coat pocket for the directions…she discovered they weren't there so she thought maybe they were left in her left pocket so she checked right there only to discover they weren't there either. She then checked both pockets again with alternating hands as she steadied the car steering wheel with her knees she remembered that her friend had said that it is two rights and one left. She took a right and was left with one right and a left. She took a left and was still left with one left and two rights. She tried two rights and was left with one left, and after trying just one left alone was left with two rights, and still she had not found her friend's house, which was starting to get a bit confusing. She decided to try a bit harder which was hard as she fought off fatigue and the traffic, and the first thing she did was reverse the right-left order, which she definitely thought was the right thing to do just then. Leaving from the corner she took a hard left, leaving two rights left, and still she was not there. A right and a left, and continuing with one more right left her not there yet either, and finally in utter bewilderment and near exasperation, she pulled off the road deciding the only decision she has left must be right, she sat back behind the wheel, took one deep breath and said "I might as well just sleep"

Naturalistic inductions

Probably the easiest way for a beginner to induce a trance in someone else is to use a naturalistic approach. A naturalistic approach involves talking about everyday trance states. As you talk to a client about everyday trance states they will be familiar so will rapidly start to enter trance. If you do this utilising hypnotic language the effects will be even greater.

It can be useful to write out direct scripts then change it to indirect. Writing what it is that you hope to achieve and how you will achieve this. Then you can go through the script changing anything that is too direct and that might not match the client's reality to something that will. For example, you may say '...as you approach that old wooden staircase...' which is direct and may not match the client's view of a staircase and change it to '...as you approach that staircase...' which is more general and so it allows the client the freedom to fit this into their model of reality.

To focus attention get the client talking about something that they are interested in. in the old day's hypnotists would tell the client what to think and what to focus on. To induce a trance you need to focus attention but it doesn't matter what you focus that attention on. That is one of the beauties of naturalistic inductions. Hypnotists used to use swinging watches, stroking, telling the client to look at a spot or a candle. Modern day hypnotists get clients to focus on issues, thoughts, comments, or even the process of their problem. One quick way to hypnotise a smoker is to ask them to tell you the process they go through when they smoke.

Utilise naturalistic phenomena. Anything can be used to achieve your goals. If you want to lead to a trance state you can use naturalistic phenomena leading to trance, like sleep, day-dreaming, a leisure activity. If you wanted to evoke a hypnotic phenomenon then you can use examples of times that thy have happened naturally like numbness – sleeping on an arm or holding snow, or amnesia – forgetting someone's name or being interrupted mid-sentence.

Creating responses this way will then come from client so they will be more powerful. It is completely different telling someone to laugh uncontrollably than reminding them of times they found themselves laughing uncontrollably, like in school in a classroom when you know you shouldn't, and the more you try to stop the laughter the more the laughter builds up, you know that feeling?

You can get the client to talk about something they enjoy doing that makes their mind wander and as they talk about it they will begin to go back into that same state of mind again. When you hypnotise someone you want to separate the conscious and unconscious mind. You can do this by confusing the conscious or marking out different messages to the conscious and unconscious mind.

Other useful ways for beginners to induce trance and do effective therapy are:

- Make someone talk about their problem without using words relating to the problem then use this to help do treatment

This can allow you to work completely metaphorically. You can use the metaphor they give for their problem and then just get them to play out the metaphor to a positive conclusion in the clients mind. This can be useful when you don't have enough information or time to work in depth with the client.

- Utilise everything don't think of anything as failure

If a client doesn't give the response that you expect then utilise what they do give you and acknowledge that what they are doing is what they need to do to achieve the desired goal.

For example:

If a client says that they can't relax enough to go into a trance, you could say 'How did you know that you needed to have a little tension there to be able to do good effective change work?'

- Time your rhythm to rhythm of clients breathing

This is probably one of the easiest ways to increase your effectiveness at altering someone's state. If you match their breathing and talk with the clients out breath you can begin to slow your breathing down and begin to slow down what you say and they will begin to relax deeper.

This is because breathing is such a fundamental part of life that if you match it you quickly begin to build rapport with the client on an unconscious level.

- Use fractionation

Fractionation is a technique developed where you take the client in and out of trance repeatedly which deepens the trance each time they go inside.

This can be done simply by asking the client to open their eyes then close their eyes again and go deeper.

Fractionation was created because hypnotists noticed that each time clients came into a session and were hypnotised they went deeper than they had done on previous sessions. It was realised that they didn't need to have a big gap between sessions, the same thing occurred if the client was repeatedly hypnotised during one session.

- Feedback what the client says as suggestions

For example:

Client: 'My left hand feels heavier than my right'

Therapist: 'Your left hand feels heavier than your right!'

By doing this you are telling the client true statements which helps to deepen their state and you are utilising on-going behaviour and comments to lead to the desired outcome.

- Take the client to the future to when they no longer have the problem and ask 'what did I do that helped you?'

The psychiatrist Milton H Erickson MD would often take clients to the future then ask what he did to help them. After he did this and they told him how he cured them he would bring them back to the present and do what they said he did to cure them.

It is a strong belief of all the top therapists in the world that people have the resources they need to heal themselves they just need guidance and assistance in accessing that healing power.

Post Hypnotic Suggestions

Post hypnotic suggestions are probably what hypnosis is most famous for and probably what causes the most controversy. Despite popular beliefs it isn't possible to make someone do something against their will with hypnosis.

I don't mean that you can't make people do things they are not prepared to do because you can. For example, it is possible to indirectly make someone stop smoking but if it went against any of the client's values or belief then it wouldn't work. The unconscious mind is normally willing to do anything that will maintain self-preservation so even if consciously the client wasn't willing to stop smoking, unconsciously they can still accept the suggestions. If the client is consciously not willing to accept the suggestions and the client recognises that suggestions are being given then they can interfere and stop the suggestions from working.

To do post hypnotic suggestions effectively you want to make sure that you prime them first. By priming the suggestions with metaphors and explanations about what you are going to do you prepare the mind for carrying out the behaviour.

After you have primed the suggestions you want to leave it a little while before you give the actual suggestion. This time is given to allow the mind to absorb the priming so that it is waiting in anticipation for the suggestion. This will increase the effectiveness when it is given.

As you wait before giving the suggestion it can be useful to deepen the client's trance and to take them into another level. For example, you could guide them down a staircase then through a door, or you could guide them along a country path then to a clearing, or simply suggest that a part of them can go to a deeper more responsive state of mind.

When you give the suggestions you want to make sure that it is worded positively saying what you want not what you don't want. So often people know what they don't want and then say that. The problem with this is that the unconscious mind doesn't understand negatives. It makes images of what is said, so if you say 'You won't have that pain when you sleep at night'. The unconscious mind will create an image of you being in pain when you are trying to sleep at night to know what it is not supposed to think about. The same thing happens if I ask you 'don't think of a pink elephant'. You have to think of a pink elephant to know what not to think about.

When you give suggestions you want to make sure that they are easy to follow. The more complicated the post-hypnotic suggestion the more chance there is that it won't be followed.

When a suggestion is followed the client will go back into the same state that they were in when the suggestion was given. That is why Hypnotherapists often give post-hypnotic suggestions to re-enter trance with a given word or phrase by the therapist because this is a quick way to re-hypnotise a client.

Say post-hypnotic suggestions three times at least, after you have done some priming and using metaphors. This helps to make sure that the suggestion is embedded in the mind. Use words like 'when' and 'as' to set post hypnotic suggestions and to link them to on-going behaviour.

Presuppositions (that will be covered later) work like post-hypnotic suggestions. As you are repeatedly presupposing specific outcomes you are setting up future responses. If the responses that are being set up are associated with a behaviour that will definitely happen then this also increases the likelihood of the suggestion being followed.

Remember to cancel any post-hypnotic suggestions that are no longer required or to make them very specific so that they will only happen at required times.

You don't want a post-hypnotic suggestion to close the eyes and go into a trance each time you hear the word NOW to be active all of the time. You want it to be limited to the right context and to a specific tonality and only be the therapist.

Remember to use:

- Embedded commands (messages marked out within sentences using a change in tonality or a gesture etc.)

- Presuppositions (using terms like as, when, after, before that all imply or presuppose that these things will happen)

- Illusionary choices (offering choices that lead to the same outcome, like saying: 'do you want to sit in this chair or that chair to go into a trance?' It doesn't matter which chair is chosen the outcome is that you will go into trance)

- Non-verbal behaviour like voice tonality and being congruent by exhibiting what you are trying to get. For example saying relax in a relaxing way etc.

Hypnotic Language Patterns

One of the most important and powerful tools for you to use to increase your ability to hypnotise others and to help others to respond positively to psychological treatment is the language that you use. Throughout the book there have been examples of hypnotic language.

Now is time to break the various language patterns down to learn them in a structured way. By giving examples in context previously you will already have a level of familiarity with some of these patterns.

Some of these patterns are more likely to be used than others. I have included some of the more complex patterns to allow those dedicated learners out there to have something to play with and expand on.

Hypnotic language is a way of communicating that leads to a response in the listener, initially at an unconscious level. Using hypnotic language is like using a special language to talk and build rapport directly with the unconscious mind. That is why they are so powerful to use for aiding change work.

Many clients of mine would have had to be in therapy for many more sessions if I tried to help them without the use of hypnosis and hypnotic language. Even if I know what they need to do to get better some people just don't respond or want to make any effort on their part. They want the therapist to do all of the work while they sit their unresponsive. It gives these people an escape route. They can say to others that they 'tried' therapy and it didn't work.

Using hypnotic language allows you to have a set of skills that will help you to talk to people's unconscious minds which means that you can create change that will seep into the conscious mind when it has happened. These skills also allow you to know minimal information about situations and problems yet speak in a meaningful way that sounds like you know more than you do.

Yes set

The first language pattern that I will cover is the most basic – the yes set

With the yes set you want to ask questions you already know the answer to.

Make sure the answers are always in agreement – yes

Example:

- You're sitting in that chair

- You've come here today to see me

These statements can only lead to a yes answer if they are true which means that you are increasing rapport, because rapport increases with agreement and understanding. It also builds up a response potential. It gets harder to disagree when you have been repeatedly in agreement.

Don't make all the answers 'yes' answers some of them should be implied yes answers

For example:

'You look like somebody who wants to get better?'

It is increasingly hard to answer no when you have answered yes to many questions. One easy way of getting yes's is to feedback what the client says. It sounds like you are clarifying but you are getting yes responses.

For example:

'So your name is…'

'…and you live at…'

Client: 'I don't know what's wrong with me'

Therapist: 'You don't know what's wrong with you…'

Reverse yes set

The reverse yes set is the same as above but always getting 'no' answers. By using a mixture of this and the yes set you can break up the questions. If you ask too many yes-set questions or reverse yes set questions the client can get suspicious at always giving the same answer.

The answers are still all agreement

- You're not standing up
- You didn't drive here this morning
- You wouldn't expect to go into trance before you were ready

Both said assuming I know they are true statements (truisms)

Tag questions

With tag questions you say the negative before they do so it encourages a yes answer.

People normally agree even if they disagree because the 'No' has already been said. It takes away their need to say no and encourages a yes response because people like things to be even so if a yes is said people don't mind saying no, but if a no is said people are more likely to say yes.

Don't use it too often or it can sound manipulative.

Use it when you want a definite yes answer.

- Will, will you not?
- Do, do you not?
- Does, does it not?
- Is, is it not?

- Can, can it not? ...etc.

Compound suggestion

Compound suggestions are suggestions where you are pacing and then leading suggestions onto each other, building on the previous sentence. (Pacing is where you match the client's model of reality and state what you know to be true for the client; leading is where you add on something extra for the client to follow even if it doesn't really connect with what is paced.)

This is usually done by starting with pacing observable truisms then leading towards the response you want. The idea is to give a statement followed by a suggestion as if they are really linked together. By giving sentences linked to previous sentences you are compounding one suggestion onto the next and so deepening the effects.

One part compounds onto the next.

Link these suggestions with 'and' or a 'pause'

For example:

'Look at that spot and I will talk to you'

(Pace and then lead)

For example:

'While you look at that spot (pacing), I will talk to you (leading)'

Use truisms or statements then lead with a suggestion or further truisms or statements

For example:

'You can hear my voice (pacing and linked to previous sentence in the last example), and you can listen to something else (leading)'

'Some sounds give us special memories (pacing, linked to the previous sentence and a truism), you can be interested to discover what images are associated with those memories (leading)'

Use a number of suggestions together one after the other linking them all to guide a client from where they are to where you want them to be.

For example:

- You can look at that spot (p) while I talk to you (l)

- you can listen (p) and you can begin to get a sense of how you will know when things start to improve in the future (l)

- I don't know which improvements will happen first (p) you can relax a little deeper as those improvements come to mind (l)

Contingent suggestion

Makes one part of the suggestion contingent on the other.

One part happens because of the other part of the suggestion. In reality the two parts don't have to really link it only has to sound like it may link.

You can link unrelated sentences and make them seem related. You usually link one part to the other with a time related term like 'as, during, while, before, after'

For example:

'Take a look at this book, as you think about what you want'

You can work from conscious to unconscious

Or from observable to non-observable

Or from reality to trance...etc

Or you can simply work from a truism then link with a statement

An example for problem solving might be:

'When you see someone smoking, you can think about how good you feel that you moved on from that old behaviour'

As with the compound suggestions you want to pace and then lead

'As you're sitting there with your legs crossed, I wonder what you're thinking'

'Don't allow the eyes to close until your unconscious mind lets you try to lift your hand'

Interspercial technique

Intersperse suggestions.

Mark out suggestions to the client.

Dissociate conscious and unconscious.

Mark out suggestions to one or the other.

You can dissociate or separate what you want.

For example:

- Creative mind
- Logical mind
- Emotional mind
- Problems-solutions

Embedded commands

These are a part of the Interspercial technique. Marking commands or suggestions as separate from the sentence with either a tonal shift or maybe by pausing before and after the command or with a gesture or movement etc... This causes a pattern that the unconscious mind picks up on and responds to.

For example:

'Some people find they...relax deeply...in the shower other people find they...drift into a dreamy state...when they are in the bath'

'I don't know whether...you will discover...that...you relax deeply...as you listen to my voice...or whether ...you will discover...that...you become more fully absorbed in your internal experience with each out breath...'

Illusory Choices

Binds

A bind is where you offer more than one choice with the same outcome. For a bind you allow the choice to be chosen. You give people illusory conscious choice. They can pick which response they want to follow. They also have the option of rejecting all choices.

For example:

'Would you like to sit in the left chair or the right chair to go into a trance'

(Implication is whichever chair you chose to sit in you agree to go into a trance)

Binds are of great use to therapists because as they appear to offer choice they make the client feel that they are in control because they are choosing while the whole time all choices only have one outcome.

Double binds

A double bind has a set outcome, you only ask for the opinion of the client. They may be right or wrong about their opinion but it doesn't effect the result. You offer more than one choice with the same outcome. A double bind can't be answered consciously.

'Do you think that left hand will get warm first or will it be the right hand'

(Implication is that one hand will get warm then the other. They can say which hand they 'think' will be the first to get warm. They could be right or wrong. All they are asked is for their opinion on what response they will give first.)

'Will you go deeper into trance with the sound of my voice, or will it be with each out breath that you take'

(Implication that they will go deeper into a trance; and that they are already in a trance. They have to wait to discover if it will be my voice or their breathing that takes them deeper.)

Outcome is to get an unconscious response

'You can forget to remember the things you forgot or remember to forget the things that you remember' (A double bind for amnesia)

'You can explore a rigidity without knowing that is there, or know that it is rigid without knowing how you discovered it' (A double bind for catalepsy)

'You can see things that are not really there or believe that they are there without being able to see them' (A double bind for hallucinations)

'In hypnosis an hour can seem like a minute as in waking time a whole minute can stretch into an hour' (A double bind for time distortion)

'You can drift into a pleasant memory and forget the future as it passes or discover yourself already in the memory curious about the future' (A double bind for regression)

'You can be aware of your hand and not know it's your hand or you can know you have a hand and not be aware of it' (A double bind for anaesthesia)

Adapted from 'The Art of Indirect Suggestion', By Stephen Brooks

Open ended suggestion

Series of choices all with the same result, any response gets the desired outcome.

For example:

'As you go into a trance I don't know whether your hand will go up or down or left or right or not move at all'

'Will your hand go up putting you in a light trance, down putting you in a deep trance or stay where it is as you go into a medium trance'

Not doing suggestions

You say what you want by saying they don't have to…

This can either give permission to do what is left (below example is staying still) or guide their attention indirectly to what you are telling them they don't have to think about.

'You don't have to move your arms, your body or your legs as you go into a trance'

Metaphors

Telling stories, anecdotes etc., either mirroring the clients situation or laying down a useful pattern, or seeding something for future work (like arm levitation being seeded by telling a story about a child in school compulsively answering questions in class and raising their hand spontaneously).

You can set up a specific emotion with a metaphor or perhaps use client's comments or metaphors for rapport or use metaphors to lay down patterns unconsciously in the client.

A story about circling a fort held by an evil invader, not letting food or water get in to the fort and not letting the invaders escape. After a short while all the invaders die could be used to fight warts, verruca's or even cancer.

Multiple tasking

Give people more than one task to do at once so that it overloads the conscious mind and they have to perform some of the task unconsciously which will initiate a trance or if they are already in a trance it will deepen the trance experience.

'As that hand becomes numb your eyelids become heavier, as your eyelids become heavier you hadn't thought about those sensations in that left foot until now, as your attention is directed to that left foot just notice how that anaesthesia is progressing in that hand, but don't let your eyes close until you are aware of how your breathing is changing so rapidly, I don't know whether that breathing will slow down in a trance like way before your eyes close or after your eyes close and you can be aware of your eyes closing without knowing that they're closed or notice that anaesthesia has developed in a profound way without you realising it'

'Whatever you do don't lose that attention you have now focused on that clock on the wall and as you look at that what does it feel like to stand out in the rain and hear and repeat in your head as I count backwards 200, 199, 198...and you don't have to be aware of that rigidity of the hand I'm holding up'

Presuppositions

Presuppositions are where you presuppose an outcome using terms like when, after, while, during, as, before, etc. They are useful for making someone think along certain lines and can also be useful for setting up ideas for the client to think about which builds up a future of having that outcome. Sometimes immediately directly presupposing can seem too intrusive or pushy. Sometimes it can be better to start a sentence in a way that sounds harmless. Like starting a sentence with the word would or starting it applying to a third party.

Some examples:

'Have you ever been in a trance before?'

'While your unconscious mind works at creating the changes that you desire you can begin to relax'

'What would it be like if you discovered that when you wake up tomorrow the old problem is gone; what will be the first thing you notice?'

'I told John 'all you have to do sit there reading this and allow yourself to relax''

Nominalisations

Using words that are none specific. They cause the client to go on an internal search for the meaning. This makes them sound meaningful to all people as they all find their own meaning. Use them regularly. They are words that the person has their own fixed meaning to.

Nominalisations are words with no fixed meaning like:

Curious, wonder, development, relaxing, explore, resources, pleasure, excitement, enjoyment, discover, fun, relax, meets your needs, satisfaction, etc...

Illustrations of the Use of Hypnotic Language Patterns

Arm levitation

You can **take some time to relax** (embedded command)... **you can let time stand still**... like a clock **stopping**...**giving you all the time in the world**... a clock can be stuck at quarter to three (hands at 9 & 3 representing real hands)... showing on the face with the motor behind being in control of those hands (metaphor for the mind controlling the arms)...whether they should be **left stuck** (embedded command) or **raise right up** (embedded command) to the twelve (clock metaphor for arm to raise all the way up)... **your ('you're') unconscious** (embedded command)...mind gets the right idea leaving that left behind (vague language, the unconscious mind will understand what it means)... **rising right up** honestly and effortlessly in front of you as that motor **move that right arm** (embedded command and metaphor to raise the right arm up and mispronounced word 'move' instead of 'moves')... as the other arm is **left stuck** right there at the 9 (embedded command & metaphor that the unconscious mind will understand)... as you can notice yourself walking **right arm up** (changing words (on – arm) & embedded command) to the clock you can become the clock with your unconscious mind becoming the motor... the **right hand can continue moving up** to the 12 even faster... as the wrong hand is **left where it is**.......

(Notice the various other language patterns like compound suggestions and presuppositions)

Selection of examples of different types of binds

'I don't know whether you will decide not to stop smoking until the end of the session or decide to stop smoking before that...'

'It's easy to forget how easy it was to remember that you smoked... while finding it hard to forget how easy it is to remember many happy memories...'

'I don't know whether your unconscious mind will keep your mouth closed if you try to smoke...or if you try to put a cigarette in to your mouth and discover that it won't open...'

'I don't know whether you will enjoy life more because you no longer smoke or whether it will be because you have cleaner lungs...'

'Will the memory that comes to mind be a motivated one or will it be a memory of high motivation...'

'Will you maintain a cleaner and healthier lifestyle to prove to others how capable you are or will it be to prove it to yourself...'

'You may get a temporary craving over the next few days... wonder whether it will be your extra energy that fills that craving or will it be that smile that is showing your pleasure you have because of your success...'

'Will you decide honestly and unconsciously to show people that you are proud of who you are or show them that you are proud of whom you have become...'

'I wonder whether you think that you will be aware of making that unconscious choice to permanently stop smoking now or whether it will just happen without your awareness...'

'There are times you can remember when you forgot what you tried to remember.. There are also times you can remember when you forgot what was in your mind only seconds ago.. Remembering that you forgot to try to remember what it was that you forgot.. Like now finding that you remember you will forget if you try to remember but knowing that you have forgotten what you didn't try to remember.. Forgetting why you're even trying when you know you will just forget everything that I have said but knowing it is not forgotten unconsciously...'

Confusion induction with arm levitation using complex language patterns

I don't know whether your eyes being shut will make your left **hand go numb** first or whether it will be the right hand... It's your right to decide which hand will be left right until the last minute as the hand that is not left can go right into a **relaxed state of numbness** leaving what's left for a little while as the one that's **numbing** goes right on spreading with the other one still left behind **spreading that numbness** right down through your body **relaxing** you as the one that's left catches **right arm up** as it becomes lighter leaving the other one right where it is with the one that's left **getting light** like a helium balloon left in a room with me right down with the other one left **lifting** reaching for the sky wanting to fly left just floating there up in the air with the other one still right where it lays with the one that's left with a mind of its own left **filling with helium** giving it that **floating, drifting, flying, relaxing** feeling which is down right **uplifting** with one right down and the other left **lifting up**. As that lightness may spread right down into the whole body or being left for the body to rapidly catch right on up as you now **drop deeply into a pleasant deep state of honest unconscious awareness**…

And finally a fun use of language patterns being used as an answer phone message. I think that it is important to practice these patterns and be creative. Use them in everyday life to turn them into something that you naturally do instinctively.

Gestures and Internal Reality

People describe a lot as if it is all in the world around them. They will use gestures, they will look at things that aren't there, they will point at things or mark things out etc...There was an interesting book by Geoffrey Beattie on gestures based on the first UK Big Brother show. Due to having 24hr footage for each day they could check observations of gestures and what was being talked about in a wider context and they could check back over footage and watch the footage over the coming weeks (in relation to what was talked about).

If you watch people's hands and gestures and physiology you will notice that they use all of these non-verbal cues acting out in the real world what is going on internally. These messages can be watched for congruence. They can be watched to see if they match what the person is saying or not. If they don't then this could be worth investigating further. Many times they convey a metaphor of what is being discussed, either matching the words or adding extra information to compliment the words. For example someone may say they have to get things in order whilst making a gesture as if marking out a row of slides one after the other. Or they may talk about bringing what they have learnt together whilst making a gesture as if gathering something up with their hands and putting it all into a ball.

If someone rubs their neck while talking about a partner that is 'a pain in the neck' the rubbing the neck is a metaphor non-verbally portrayed, as 'pain in the neck' metaphor.

Likewise if someone digs there heals into the ground as you talk to them about changing people say 'they were digging there heals in they didn't really want to change' so non-verbally they convey this as a physical metaphor...

Working with parents I have watched them talk about their child whilst doing wringing actions with their hands ('wring his neck')

There are many such metaphors that can be observed played out using nonverbal communication

If you imagine that you are in the clients' internal reality then you can interact with it, another technique you can use is spatial anchoring. This is where you mark out different anchors in the space around you. For example if you do this with a group it could be that every time you say or do something you know will make people laugh you gesture with the right hand out in front of you at shoulder height; then when you create a state in the audience of curiosity you gesture with the right hand at waist height; then perhaps you talk about learning effectively and you gesture up by your head, etc...

As long as you know what gesture is for what response you can then use these gestures to make the person re-enter that state. So if you want them to access a state of curiosity later in the lecture you set off the curiosity anchor. If you want them to access humour you set off this anchor, if you want them to learn something easier then set off this anchor.

You can also push people's pictures forwards when talking about associating or getting a closer look, and gesture moving pictures away when making things more distant or disassociating. It is important to be in the clients world and interact congruently with it. By that I mean that if you want a client to look at a hallucinated TV screen then you look at it as well. If you want the client to see something closer then gesture moving it closer.

Erickson's Early Learning Set

One of the main reasons that Milton Erickson used the early learning set is because it triggers state dependant memory learning and behaviour. In therapy he would want to have the client in a state of mind where they can learn effectively to implement the therapy.

Priming research has shown that if someone is asked to think about old people they move slower, think about young people they move faster, think about depression they feel move down, think about happiness they feel more happy. If a therapist talks about family relationships the client thinks about their own family relationships.

Erickson would often discuss learning, how you don't remember how you learnt but you have the results so obviously you did learn he would go into detail (demonstrating how much was actually learnt) and he would want to provoke this mental set in the client. If this gets provoked then the client will be in a mental set where mistakes are a part of learning (like falling over hundreds of times before finally walking correctly), where complex tasks can be integrated into who they are and can become something that goes on outside of awareness once learnt, the person will only really be aware of the results.

Hypnosis & Trance

I am fascinated with 'designer trances' that all emotional states are trance states and that as everything has a level of emotion in it everything has a level of trance involved. I believe that you can mix different trance states to create new states (like mixing jelly belly beans to get different flavours, or mixing cocktails). I use musical rhythms etc to create altered states and by starting one trance then adding another I create designer trances. This is what I do in many of my audio tracks with music (I'll even put in nursery rhymes etc as these cause trance states). Because everything has a trance element the problem is the easiest thing to use to hypnotise someone. I remember being told on a course many years ago that if you want to hypnotise a smoker ask them to describe smoking and they will enter a 'smoking trance'.

My view on hypnosis is also that hypnosis is actually the art of inducing different trance states not the induction of one state. And that hypnosis is just the term given to the advanced communication skills used for doing this. My opinion is that the classical 'hypnotic' trance state is actually where a person has been guided into a 'peak learning state' where they are able to learn new information (updating old patterns) or learn greater control over themselves (like over unconscious processes etc.). Hypnotic techniques can also be used to induce relaxation, anger, confusion, fear, pain, love, sadness, desire, etc...

Because you can induce all of these different states having the 'state vs. non state' argument to me seems non-existent as it seems like an argument over something that is in effect nothing but a collection of advanced communication skills techniques on the part of the therapist (or in the case of self-hypnosis someone being skilled enough to induce a desired state of mind without external intervention). I think when it is induced indirectly it is hard to think of the induced state and any hypnotic behaviours as being 'acted' if the person didn't know they were being hypnotised or what is being expected of them. Whereas when it is induced directly and they are told what to do some people may fake it really well.

The brain is essentially a pattern matching machine for survival. These patterns can be added to and updated all of the time.

This process happens all of the time, if you have a phobia you see something which sets off a feeling of fear (often making you run or freeze before you think what you are doing) then you start to think.

ALL hypnotic language patterns (I'll call all things that can be done or used to induce hypnosis in that definition verbal and non-verbal) are recognised by the brain unconsciously and the pattern is understood leading to a response (If the pattern isn't understood there is no response) So in effect you are able to create responses with the person essentially in any state, not confined to a 'hypnotic' state. But unlike most arguments for the non-state idea the person has no idea why the response happened, it wasn't expected of them so they had no idea to play along consciously...it just happened.

I believe that trance states exist. My definition of a trance state is a narrowed state of attention on a stimulus that defines your behaviour in a way that will aim to maintain that state (I also believe that a complete lack of attention like in meditation is also a trance state because (if it makes sense) they are focusing on one thing – not attaching to anything)

You get many natural trances (including what could be termed a hypnotic trance) Anger is a trance state, so is love or relaxation etc... I believe that the stronger the emotion generated the deeper (as it were) the trance is because it is harder to break free from it and overcome it. Recent research (for more information visit www.humangivens.com) shows that a 'hypnotic' trance is an accessing of the REM state. This is the state of mind we enter to update patterns of behaviour. This doesn't mean it is necessary to put someone in that state to get phenomena though, it is just the state we update patterns and learn in.

ALL 'inductions' use processes that cause this state of mind. Shock or confusion trigger the re-orientation response which is the REM state to lock on and learn how it should respond to this unknown situation, Relaxation, guided imagery etc. are all parts of going into the REM state. As is getting the eyes to move side to side.

I see it as a selection of natural processes that are being utilised. The hypnotic state is good for updating patterns of behaviour so it is a state, but recognising patterns and responding to them is non-state specific.

Working with Ideo-Dynamics

You get a range of different classes of Ideo-dynamic responses

- Ideo-sensory: A sensation that can be noticed by the client

- Ideo-cognitions: Thoughts or images coming to the clients mind

- Ideo-affective: The client experiencing feelings

- Ideo-motor: The client experiencing an automatic movement

If I want to notice a response in a client but also want to give the client wider choice I'll ask for clear signals I can notice and then pay attention. Or if I want to offer even greater freedom I will ask the client to tell me when the unconscious give a signal for yes and to tell me what it is, and the same with no.

This way the client is a bit more involved and so it isn't appropriate for all situations but some clients will happily say it's a feeling in their stomach, or a shiver, or a warmth in a hand or foot (ideo-sensory) or a voice in their mind or a word or an image in their mind etc. (ideo-cognitions), or a feeling of comfort, anger, sadness, happiness etc. (ideo-affective), or a movement in the toe, finger, twitch in the face, twitch in the leg etc. (ideo-motor)

When using parts for therapeutic interventions often feelings can be used, where you can ask for different parts to be there and establish with the client what those parts are and how they are expressing themselves (I'm a strong believer in allowing self-expression) and have people notice the sensations as the parts integrate into the new learning/understanding/resolution etc...And they can tell you when this is done.

Ideo-motor movement could be done in a similar way with (for example) a hand representing the problem, and a hand representing the clients resources and have them move (self-expression again) until resolution is found and the problem and resources have integrated into something new.

The above is using these responses in an open way rather than yes/no.

When I have asked people to notice a signal that means yes and a signal that means no and to let me know what they are it works in the same way as finger signals except if it is something you can't see you wait for the client to get the answer from their unconscious then they give you the answer verbally.

Often when you are being observant they don't need to give you the answer verbally as you will notice the ideo-motor movement of the head nodding or shaking as the thought begins to seep through just before they answer.

So in effect you will still be reading ideo-motor signals but you have set up (in a way that suits the client) a yes/no signal set that the signal will come from. If for example it was warmth in the left hand for yes and right hand for no when the answer comes through you will notice a hand getting slightly more red, then you notice a fraction of a second later a slight movement of the head yes or no, then movements associated with being about to speak, then they tell you.

Hopefully all of this will be congruent, but if it isn't the chances are the bit that isn't would be the conscious verbal answer.

Over the years I've not really noticed any one type of response to be better than any other except that sometimes people do things too consciously and seem too consciously involved and if you are time limited it is easier to have a yes signal without saying what it is to be and watch for it and the same with the no, so that they don't just lift the finger consciously because you were setting up finger signals and they knew that and so thought that would be what you would want.

I like minimal input in words, just getting agreement from the unconscious and observing signals as the unconscious does the work (videos of Ernest Rossi doing this are very good and informative), then a signal to say the work is either finished or is now at a point it can continue all by itself.

There are many techniques and schools of therapy that use these other responses, like various 'parts' techniques and therapies where the client may be ask to call up a part and wait for the response and the say what that response is/where that part is - like a feeling in the stomach, or tingling in an arm etc...

Subliminal Auditory Stimulation

What is it?

Milton Erickson discovered that he could influence people by matching then leading people's breathing patterns. He called this Subliminal Auditory Stimulation. From his studies into this he developed this as an important part of the way he worked...

It is interesting how using this once rapport has been built you can create thoughts and words in others that they think they came up with themselves.

I have often wondered how many apparent psychic examples can be attributed to this process, either done with intent or done without the operators' knowledge.

It is a two-way process

If you go with it you can let messages come through from the other person, especially if you allow a trance to develop first. So what you would be doing is effectively having the client breathe what is in their mind and you begin to breathe the same, rather than you initiating it by breathing something for them to pick up on.

Back in the 1960's Erickson wrote a report on experiments he carried out on influencing people just by breathing.

He termed this 'Subliminal Auditory Stimulation'

Milton Erickson would sit next to, or in view of the person he wanted to manipulate. He would then breathe the same as them for a while before changing his breathing. This change in breathing would then also happen in the subject. He tested this by triggering stuttering, yawning and humming or singing songs.

Over the years Milton Erickson refined his ability to influence people with his breathing technique. He would use it to hypnotise people, to make people fall asleep and to influence what thoughts people have.

Since the 1990's when I found out about the work of Milton Erickson I have used 'Subliminal Auditory Stimulation' to influence others. I have used it in Business Meetings to get others thinking what I want them to say, I have used it in meetings and in public situations to make people fall asleep, I have used it in childcare to get children to sleep, in therapy to influence my clients decisions and I have used it regularly to induce a hypnotic trance in people.

So far, despite having success on many occasions at using this technique I don't know of anyone that has carried out a proper study on the subject. Without carrying out a full scale study there is no evidence to support the claim that you can influence people by breathing in specific ways.

As well as demonstrating that it is possible to influence people consciously in this way, it may also lead to explaining some claims of psychic abilities. For example: when one partner is thinking about wanting a cup of tea and the other partner gets up and offers to make one, or when you get those moments where a friend says something and you discover that you were thinking the same. It may also explain why some people feel they are psychic, because they naturally have an ability to unconsciously pick up on the breathing patterns of clients.

What I hope to do is to have as many people as possible carrying out the experiment in real situations. These people will choose a target; sit next to, or in sight of the target. They will then spend a few minutes matching the targets breathing before changing their own breathing to breathing a tune. This is a tune that should be recognisable to most people. They will 'breathe' this tune for a few minutes, before recording the results.

What they will be looking for is how long they spent matching the targets breathing before they were able to lead the breathing. Then how long they spent 'breathing' the tune. Then how many people responded by humming or singing that tune.

Also people to use it in daily life like matching breathing then asking in your mind for a cup of tea, or for someone to say a specific sentence.

Anyone reading this that decides to give this a go I would be curious to hear your feedback.

Learning to Notice Minimal Cues

Over the years I have studied many martial arts, one of my favourite 'party tricks' used to be grabbing a coin from someone's hand before they close their hand. I would do this by watching for the first minute movements to indicate that the arm will be moving and the hand will be closing. The same with noticing punches etc... The interesting thing about the Wing Chun practice of 'sticky hands' is it is easier blindfolded as all of your attention is on what you feel with fewer distractions.

The best way to practice noticing minimal cues is to make this part of your life. Practice observing people when you are out anywhere other people are. It could be watching people in restaurants, or it could be on a bus or in a park. Even on TV and watching programmes like Big Brother.

It is useful to limit what you are looking for rather than attempting to see everything all at once. Or limiting to watching a specific area - like the eyes, or the mouth etc...

As you watch people look for patterns.

You can also practice with friends. Get with friends in pairs and you can do some of these exercises:

1. Sit opposite each other, ask the other person to think of something they really like, then change the subject a few moments, then ask them to think of something they don't like.

Watch them and ask them to think of one then the other slowly a few times, then to randomly think of one or the other and you tell which they are thinking of. Do this a number of times and notice what you are noticing that lets you know.

Then do the same again but this time sit back to back and have them count 1-10 while they are thinking of one then the other and then have them randomly think of one or the other and you work out which from their voice.

Then do the same again with your eyes closed and the palm of one of your hands touching the palm of one of their hands. Have them go through thinking of one then the other then randomly thinking of one or the other. You work out which they are thinking about from the kinaesthetics.

2. Have a friend sit opposite you and think of 2 truths and a lie. Notice what is different about the lie

3. Have a selection of different coins, practice noticing subtle differences by hearing (with eyes closed or back turned) different coins being dropped one at a time and say which coin is being dropped (on a table is best)

Hypnotic Language Patterns, Skills and Ideas for Working with People

Open-ended Suggestions

As you look at your hands I wonder whether you will notice the movement that will occur as you enter into hypnosis? Will the movement be small twitches or larger movements... or will it be a lifting or sliding... or pushing down... or will it seem to go unnoticed... and seem to be incredibly still... and will that movement be in the left hand... or the right hand... or both hands...and I wonder whether it will start in a finger... or in the palm of a hand... or perhaps in the back of a hand...or if the movement will start from elsewhere in an arm to create that movement...you can be curious to discover how your unconscious expresses your own unique way of entering into a state of hypnosis...etc...

Open ended suggestion allows many options (almost anything can happen) yet there is only one outcome making them similar to the various forms of binds...

Nominalisations

Utilising Negative Nominalisations

Negative nominalisations can be used as a way of describing the problem even if you don't know all the details, if you use the term anxiety the client will know what you are talking about even though you may not really know what it means to them.

If you then talk whilst leaning to look behind the client whilst mentioning anxiety (for example) and put it in past tense you can place the problem in the past, you can mention 'back there' etc...

You can also alter the meaning of a nominalisation, so you could start with 'anxiety' and their meaning and begin to reframe and alter the meaning of the nominalisation so that when they think of 'anxiety' it has a different meaning to them now to the meaning it had before the session.

A Collection of Positive Nominalisations & Ambiguous Statements

Express the true you

Discover the qualities that make up who you are

See yourself as if through the eyes of someone that loves you dearly and discover what comes to mind

Voyage of self-discovery

Discovering your sense of purpose and meaning in life

Following your heart

Self-realisation

Self-discovery

Authority

You can use your leadership skills...

...Feels like to become top dog...

...Display your management qualities...

...Get a sense of what it feels like to be the leader of the pack...

...notice how things can go your own way, even turning bumps in the road into opportunities for success and achievement...

...Use your strengths to achieve success...

Charm & charisma

Discovering a feeling of respect from others

A feeling of love and friendship

Connection with people you meet

Feel that special feeling inside that lets you know those around you care

Part of the community

Feeling of togetherness

Meaningful relationships with friends and family

A binding sense of unity

Using Guided Imagery Journeys

Hypnotic journeys can be used with adults and children alike. Most people enjoy a journey or adventure. With a guided journey each change of scene create a new depth of trance so for example; if someone walks along a beach then walks into a beach house they will now be in a second level of trance, if they then settle down in a comfy chair and relax and drift into a dream they will now be in a third depth of trance.

To bring the person out again you need to reverse the route they took. This sandwiches the deepest part and each subsequent layer so that on coming out of trance the client will normally have considerable amnesia for the deeper parts.

Some examples of journeys:

Walking through fields towards some woods in the distance...then walking through the woods following a stream up towards some distant mountains, then climbing the mountains to find a cave...and in the cave you start a small fire and look at the cave paintings that appear to flicker and dance to the light of the fire...getting a sense of becoming the spirit of an Eagle leaving the cave and flying over the land below, over the route you have just taken, getting a whole new perspective, noticing how everything can seem so still from up here...etc...

Or one of my favourite journeys to use for inductions (when I use more structured inductions) is to guide someone through a woods to the edge of a vast ocean (as a land animal), then under the water becoming a whale or similar, then swimming down to a cave entrance deep under the sea that gets followed and comes out in a secret land that is like it has been set inside a mountain, with a forest and a house in the middle of that land, and that house containing unknown knowledge...(spend sometime in the house before reversing the journey)

Or I talk about a prince that looks out from his castle to see people suffering, wondering why he should get what he wants and never suffer yet all these people seem to suffer so much, he gets in disguise and goes out to see what it beyond the castle walls. He walks through fields, meets peasants that teach him, he goes throughout his land and eventually settles under a tree to meditate on what he has learnt; he goes on that mental journey before making his learning's then heading back through his land to the castle to share and use what he has discovered...

Embedded Commands

Embedded commands are used positively and negatively in everyday situations

For example:

Doctors or dentists telling you 'this will hurt'

People telling children 'you're never going to amount to anything' , 'you're rubbish at maths' , 'one of these days you're going to get hurt doing that' etc...

Doctors telling you 'the problem will last for three to six weeks'

To children 'you're going to be so successful when you grow up'

Using Ambiguous Language, Mispronounced Words and Confusion

Don't let a trance state develop until I have counted all the way down from five... and you can respond to any form of counting... and notice how it develops in stages... with each count from five... and before (four) you three (free) yourself... and allow yourself to (two) go all the way in trance... you can discover what it means being at one now... with your thoughts and feelings and a deep sense of relaxation... as you notice that developing in its own special way... wondering as you wander comfortably... drifting and relaxing... whether you will go deeper with the words that I say or the spaces between the words...(then at the end of the induction)...and as you drift back to a sense of oneness you can rise up to (two) an awakeful state with each count...feeling a sense of threedom (freedom) and before (four) you come all the way back you can let go of the experience on a conscious level and end the counting on five...then opening your eyes...

As you wander along wondering where the wandering will take you, you can wonder what else you will discover as I talk to ...your unconscious (a command to be unconscious/in trance)...mind, it is your right to decide whether your right (correct or right) hand will lift up or your left hand will lift up, and you know which hand is right (correct or right) and the right (correct or right) hand can lift while the wrong hand will be left ...(left hand is wrong/wrong hand will stay where it is) where it is, and you came here today and noticed that it is a nice day for a change (good outside, good day to change)...not like the other day where the weather just makes you wonder whether ...it is worth the change (command that it is worth changing and money change) ... you spend on tents (tense) when it takes all that effort to put them up and tents (tense) come down so easily and effortlessly, and you can take the tents (tense) down in so many ways it's impossible not to be able to take the tents (tense) down...and as you glisten (listen - Commonly used by Bandler) up to each word I say and be calm (become) aware of what it is like to wear something different and try on something new and wonder where you'll wear that...in what situations and what contexts...and in a minute you can take an (on) hour (our) ... discussion and the meaning and discoveries you have made and discoveries you don't know that you have made and wander (wonder) through what's new and realise that 70% of discomfort is made up of comfort (discomfort = 10 letters, comfort = 7 letters) and I wonder what that will mean to you the next time someone is mean to you and you take their meaning and pick up what they mean in a new way a way you didn't know you knew...dismissing the dis (common youth term - dissing you - meaning putting you down or being mean to you) and discovering (diss covering - hiding it) the comfort in yourself...

I enjoy mispronouncing words that most people seem to overlook or not notice

'You can take every trance (chance) you get to really relax'

'I don't know if that movement will be in the lift (left) hand or the right hand'

'I don't know which hand will lift right arm (on) up into the air'

No more/know more

insecurity/in security

Nowhere/Know where

Tents/Tense

Wonder/Wander

Stairs/Stares

In trance/Entrance

Heal/Heel

Changing state, or going from state to state etc... (Driving down USA for example - changing states of mind)

Breathing and Minimal Cues for Deep Rapport Building

I match the breathing from the start of the session and keep it matching. I won't match bad breathing patterns like coughing or an asthma attack etc., but normal breathing I match.

I will talk on the out breaths, unless I am going for an arm levitation when I give suggestions on the in breath. If someone is breathing too fast for me to be comfortable with I match every two breaths so that I do one whole in and out breath to every two of theirs.

I believe matching breathing is one of the most important things to match as it is such a fundamental process and it builds a really deep connection.

After matching breathing for some time I lead by deepening my breathing and lowering it down to my stomach as they now follow my lead.

I rarely stop matching the breathing (there are occasions I do but not often), I'll talk in time with their breathing, breathe in time with their talking, and breathe in time with their breathing.

It can be practiced in any situation where people are present. Go to public places and you can practice matching breathing of people there and also even more usefully you can practice noticing peoples breathing. With people that breathe really slowly it can be difficult to notice at first so practicing on hundreds of people every day really helps. You can then put people in different states just from matching (pacing) then leading their breathing...This can be fun to do with people in libraries, on benches, on buses or train journeys etc...

I remember an experience many years ago when a hypnotherapist said he couldn't enjoy being hypnotised by indirect hypnosis as his knowledge was such that he would notice what was going on and so his experience wasn't like the experience of trance spontaneously appearing to develop out of nowhere like his clients would have and like he used to have when he started out. He didn't believe it would be possible to experience that again.

I took this as a challenge. We sat discussing what he was saying and then carried on into 'ordinary conversation'.

As we were talking I kept asking questions I knew would make him go inside his mind, but in context with what we were discussing. I could see his pulse in the side of his neck so I matched this with the movement of one of my feet. I matched his blinking with one of my fingers whilst matching (on the surface) his general body posture and hand positions etc...so to him I would have appeared to be matching him and it was the subtle information I was cross matching (of which he was unaware). Just like when you pat your head and rub your belly, to do this I aligned myself with one thing at a time, moving on to add more once I was comfortable with what I had got.

I matched his breathing normally just matching it with my breathing, I matched his external/internal focus with my overall body posture (sitting taller when he was looking at me and talking to me, relaxing my muscles and slumping when he went inside his mind)

It didn't take long for him to be at a point where he clearly wanted to close his eyes and go into a trance but as we weren't in a context where we were doing hypnosis he would have been closing his eyes in the middle of what was appearing to be an ordinary conversation. He clearly was waiting to have permission to close his eyes. I gave him this permission, he went very deeply into a trance and when he came out of the trance he said that it was the deepest trance he thinks he has been into.

How to Match and Mirror Successfully

I remember reading something Derren Brown said once, if you want to match and mirror people properly before you meet them get in the mind-set that you like them as much as your best friend. Then when you actually meet keep hold of that feeling that you have known them for years like a long term best friend and you will fall into rapport with them naturally which then won't look faked.

I tried this a few times and found that for planned meetings it worked well (times when you can get into state first). I found that from videoing dozens of interactions with people (me with others) when matching and mirroring was natural overt movements had more of a delay than subtle movements. Things like shifting in the seat would take a second or two, gestures would only match if the context was the same (so my gesture would hold the same meaning), things like angle of the head, leaning etc. matched pretty quick, matching the types of words and sentence structures and tonality etc. seemed immediate (next thing that was said) unless it didn't fit with what was being said (like if telling story or putting on a voice etc.) Breathing seemed to match rapidly as well and so did heart rate (either seen in the neck or if the legs are crossed, seen in the movement of the foot that is off the ground, or in the wrist by the thumb etc.). My assumption was that to fake it I had to apply this and match more minimal cues quicker than more overt cues and that I had to make sure what I was doing also match meaning (so if someone gestures throwing a problem over their shoulder I can use the same gesture when describing getting rid of the problem so that it shows a deeper understanding rather than just copying a movement, likewise if someone demonstrates a churning motion with their hands when they talk about their problem and how it feels I don't just copy it I use it in context when talking about that feeling).

Some ways I have practiced matching and mirroring is using it whilst sitting on buses and trains, matching other passengers and then leading them into a trance state.

I think matching breathing is one of the best ways to deepen rapport. Many courses teach to match body posture, gestures, clear movements etc., what often happens is that people look like they are copying the person and it can make the person feel uncomfortable. Whereas breathing often goes unnoticed. My view is that the more obvious something is the more careful you have to be in matching or mirroring it. I think that the best way to do it with more obvious movements is to make the movement or change when it is appropriate to do so. So if someone is sat with their legs apart then crosses their legs I wouldn't immediately copy this, I would wait until it is appropriate to do so in context with what I am doing and saying. I also wouldn't do any movement that was unnatural or uncomfortable for me to do; in these cases I may do cross matching, so if they cross their arms I'd cross my legs.

Using matching and mirroring is a good way to get people to talk with you. For example if you see someone you like before you talk with them (say in a bar) you can match them, they will pick up on this and begin to feel a connection to you even if they aren't directly paying attention to you, they will feel like they know you or like you but won't know why or where from.

In therapy situations by becoming as similar to the client as possible you can begin to get a sense of what they feel which can help your understanding of their problem on a deeper level, rather than just what they are saying.

My Friend John Technique

The My Friend John technique is a way of hypnotising somebody whilst appearing to be talking about hypnotising someone else. There is an example of Erickson doing this to an interviewer:

The interviewer asks (as you never should to a hypnotist) 'How do you hypnotise someone?'

Erickson replies with (can't remember the exact wording but I'm sure you'll get the point):

'Well firstly I look at them like this...and I say "I would like to have you pay attention to the words that I am saying...and as you pay attention to the words that I'm saying you can notice how your breathing is slowing down comfortably...how your blink reflex is slowing down, how that immobility is setting in all by its self"...etc......

Watching the interviewer nodding in agreement as he drops deeper into a trance without even knowing what is happening is great.

Putting Yourself in an Externally Focused Trance

In one of the Erickson/Rossi books; Erickson is asked about the state he goes into when he is working with clients.

He explains that if he thinks he is likely to miss something important he will begin to pay very close attention to minimal cues, first starting with one cue (say movements around the eyes) then adding another cue (say pulse rate in the temple or corner of the eyes) then after a short while add in another minimal cue (say colouring of the cheeks and cheek muscle tonus) then would add in another minimal cue (say lips - blood flow to and from them etc.) etc...

He said that as he does this his attention becomes increasingly focused on the client and he enters an externally focused trance state where his conscious is like an observer and his unconscious is doing the work. His unconscious is noticing the minimal cues and patterns and using them without conscious interference.

I have found this an excellent way of inducing an externally focused trance state to enhance therapeutic ability and from noticing patterns whilst keeping track of what is being said you begin to notice almost like a second dialogue that is running parallel to the conscious dialogue that is based on ideas, concepts, patterns and unconscious self-expression.

I used to do this when playing pool, I would become aware of my breathing, my heart beating, the feeling of the weight of my arms, of my hands, of the balls moving, the sound of contact, then when it was my turn I would continue this to include the feeling of the steps around the table, the movement of the cue etc...and 'it' would play, 'I' would observe...

Peripheral Vision

Peripheral vision is far more capable at detecting movement so if it is used in observing people you notice micro movements easier.

It is also the vision you use to do photo reading or rapid observation when you want to detect movement and take in more information rather than colour and fine detail. If you watch someone with peripheral vision you can be looking at their face and see movement of the hands or legs and breathing, and easier to notice the pulse in the neck or ankle or wrist etc whilst noticing colour changing and overt movements and responses as you are close enough to notice these. You can also notice other people if you have a group around you even though it appears you are not paying attention to others around you. And likewise if you are talking to others you can keep an eye on the subject (or targeted person if in a meeting or social situation, like dating when you want to watch someone's reaction and responses)

Not Doing to Create Change

I remember a session of Milton Erickson's where he hypnotises a client and then leaves the room. Later he comes back and brings the person round.

The person says: (not word for word but you'll get the idea)

'I don't remember you saying or doing anything?'

Erickson: 'You don't remember me saying or doing anything'

Client: 'No, but you must have done something'

Erickson: 'I know, I must have done something'

Client: 'If you did I don't remember what you did'

Erickson: 'You don't remember what I did'

Nothing was done but the person changed, they knew how to and what was needed.

I have done this once on someone that if I verbalised anything they had a 'but' for it, or a counter argument, even if I was only saying what they had said. They were very extreme in this behaviour. The idea came to me that I could gather information opening relevant and useful patterns and associations in the clients mind related to resolving and reframing the problem, their unconscious would hopefully notice what I am doing. I then explained that during hypnosis some people hear what is going on others go so deeply into a trance that they don't hear anything. I explained that all relevant ideas and suggestions that will lead to healing will go unheard and that his unconscious will understand and know all that is needed from this whole session... (Obviously I said more than just this and then did the induction) after a while and also doing the suggestion 'I don't know if it will be the words or the spaces between my words that will help you to go deeper and deeper into a comfortable stable relaxed trance state'. I spoke less and less and what I said was not full sentences to try to mimic (for the client) occasionally rising out of trance a little hearing a random bit then lowering, then I shut up (didn't have the confidence to risk leaving the room, my plan was if he opened his eyes I would respond as if I had just counted him out). After about 20 minutes I started talking again (initially with fragmented sentences) and brought him out of trance mentioning that his unconscious can use all of the information gathered throughout the session to help the client, and then I gave illusory time frames as a double bind. Then I just utilised his responses playing on the apparent amnesia he thought he had and he left the therapy centre and I saw him in town a few months later much happier and more engaging in life than he had been. Yet I had done almost nothing, he did all the work and I don't even know what that work was.

The 'Just Being There' Trance Induction

The 'Just being there' trance induction is something that I believe can be done, in my experience it isn't that you are doing 'nothing', you will always be doing something or the client will have some expectation. I now walk into situations and people will have said I can hypnotise people just by them being around me as if I emanate a 'hypnotic force'. I can just walk up to someone and they will 'expect' to go into trance and so do.

When you are intensely focused on someone and matching/cross matching breathing, body posture, pulse rate, blinking etc, and then you lead them into a trance it can look like you were doing nothing yet you were doing something.

Doing nothing can also apply to sports and other activities if the 'doing nothing' is consciously doing nothing. If you try to play a musical instrument you do better when you stop thinking about it, if you try to play pool or snooker you do better when you don't think about your stance, the movement of the back arm, whether it is straight and relaxed etc..., same with most sports...even walking, everyone has probably experienced suddenly thinking about your walking and suddenly feeling uncomfortable when trying to walk...My view on this is that all these things once learnt (even me typing this now) are done much faster and more efficiently when done unconsciously than when you are trying to focus on doing everything at once consciously, so letting go improves your abilities. Just like letting go improves the ability to visualise rather than trying to visualise or your ability to have more productive thoughts, or your ability to relax etc...

Relaxation; Trance and Trance Signs

Soldiers go into a trance marching and are completely not relaxed.

I have worked with people with things like phobias in the situation they are having the phobia where there is no way of relaxing them or doing formal hypnosis of any sort, because the person could be at the top of a tower about to abseil or about to hang onto a death slide; so they need to be helped in minutes.

Often people are deeper than they first seem, and when you watch them you notice the trance indicators even though they seem alert and awake...

An induction I like using is to guide people deeper into a trance by using a painting in an art gallery in their mind; or a room with a TV or Cinema screen they can step into followed by other paintings or screens in those paintings or screen that can take them deeper and more fully absorbed.

Another induction that is less structured word wise and offers more freedom for creativity for the therapist with the sensory language input is to use a journey or adventure with multiple routes to take. You could use a holodeck or the adventures of Alice in wonderland with doors and rabbit holes, twists and turns etc all leading to new discoveries. I have used people settling in a situation like by a flickering fire and drifting into a meditation where they can find themselves in an experience where they see themselves meditating under a tree perhaps, then they can lower into that them and discover where that them is meditating about etc...I think it is useful noticing processes not just taking a technique and using it as it has been learnt.

I feel the more out of the process the therapist keeps the better, so unless a client mentioned holodeck or star trek I wouldn't do that, I wouldn't tell them where they are meditating etc...

In my experience you don't need the person to appear to be in a deep trance to have effective therapeutic work take place. I have had people either too hot or too cold or uncomfortable and they have moved or opened their eyes and got sorted out without any real negative effect on the trance or the work done.

People go in and out of trance with each sense system. At one point they may hear a noise so their hearing comes out of trance whilst the rest of the person stays in trance, then when they are comfortable with the noise their hearing goes back into a trance again, then if they need to open their eyes their eyes come out of trance yet their hearing and feelings are still in trance, then they go back again when ready. Likewise if they are uncomfortable how they are sitting they may come out of trance to shift position then back into trance again.

I have experienced this in many situations, even when people have had to get up completely because they hadn't turned off their phone and it has rung so they have come round, answered it and then afterwards sat back down and as soon as I have continued talking in the same hypnotic way they re-enter trance almost instantly.

When I started out I used to worry that people had been asleep, not hypnotised. I had regularly been reassured that it is incredibly rare and that in the rare occasion that it happens it usually mean the person needs the sleep more than the therapy on offer. Nearly every person I hypnotised told me they were asleep because they don't remember anything. I learnt that they weren't because they always opened their eyes on cue.

It became apparent that these people's unconscious was listening and that the conscious had gone off somewhere else...The unconscious always responded in a way that let me know it was listening. I used to test if I was unsure by getting the client blushing on one side of their face (as a visual cue not easy to fake that if they are sleeping they are unlikely to spontaneously do)

I used to get hung up on having to get people in a deep trance. Now I am so unconcerned as I know the unconscious is always listening, most of what I do probably doesn't resemble hypnosis. This is something Steve Gilligan said when talking about doing

Deep Trance Identification as Milton Erickson, he said that one thing he learnt was that everybody's unconscious was listening and that everybody was already in a trance.

Creating Dissociation, Metaphors and Age Regression

My view is that a metaphor is a form of dissociation. The metaphor maybe used to give the client a pattern for the necessary association, for example if I told someone about streams running to a river then the river running to an ocean it would lay down a pattern for association or integration, the result of the metaphor would be association. Likewise if I told a story about a cat that lost its kitten and blamed herself so couldn't cry or express emotion and spoke about the cat's experience of getting in touch with the feelings, that would be laying down a pattern and would be dissociated from the person and their actual situation, hopefully getting in touch with those feelings would happen as a result of the metaphor.

Likewise with smokers I would often discuss paths in forests and cutting through a new path that once cut through sufficiently is easier, quicker and more pleasurable to follow. Or about someone moving from a smog filled city, with all the congestion on the roads etc... To moving to the country with open space, fresh air etc...

Even metaphorical tasks have the person dissociated as they are experiencing a pattern (like going into a field to find two identical blades of grass) that will be of use to solving the problem, they aren't actually experiencing the problem itself, or the solution itself.

Using the Crystal Ball Technique

I had a client I used this technique with; again rather than the standard version I created my own version based on the principles.

I started by talking about how the sun is 8 minutes away and how what you are actually seeing is what the sun looked like in the past and how you can't see what is happening now. I then mentioned other planets and stars and how each of these is also being viewed at different times in the past, that when you look at the night sky you can see many stars and planets and the moon simultaneously yet the moon is what it was like seconds ago, the planets are what they were like minutes ago, the stars are what they were like years ago...

I then moved on to talking about how as a child you get so engrossed and focused on cutting exactly around the shape of a person on folded up paper, and that the more you cut the more careful you are to make sure you cut it really well, and you focus completely on that paper and on that cutting...you can't wait to see what it will look like...you're excited to find out...when you have cut out the shape of a person completely you then slowly and carefully open up the paper to have it look like lots of people all holding hands...

I then mentioned how each person can carefully be coloured in so that each one looks slightly different from the next, with a youngest one on the left and the oldest one on the right...and how each one can be coloured in to represent a relevant part of how the issue we discussed (it was smoking) was able to form, formed and had been able to be maintained up until now...

Then I spoke about how many great works of art have many layers where they have been changed, updated and corrected until the artist feels that the picture is right, and that they will take all the time they need to make it just right...

I told them they can continue to slowly and carefully colour in each person until each person feels just right...and how they can get a sense of how colouring in one person can influence how the others need to be coloured in and altered...

I had them do this until they were proud of their work and could step back and admire the end result (head nod to let me know this was done), then I had them carefully fold back together the paper noticing how it can become more 3D as each part is stuck back in place with the newly painted images integrating in their own unique way...

I moved on to some more stories before moving back to talking about space, stars and planets then the sun, and then allowing them to open their eyes when they have fully reintegrated in anyway necessary and made all the changes needed to allow progress to be automatic and to take effect at an appropriate rate and speed...

Another way for inducing age regression can be to use Double Dissociation Double Binds:

'You can drift into a pleasant memory and wonder what the future will hold, or discover yourself already in the memory curious about the future'

'You can experience a pleasant memory with no awareness of the future, or be absorbed in a memory looking forward to the future'

'That memory can take you back to a previous pleasant experience before the future happened, or that memory can take you deeper into the past curious to discover the future'

When I was out in Dubai recently I found that, being from the UK, I wasn't used to all of the heat. I would be beside the pool, lying in the sun and after a while I would get used to it even though at times it was uncomfortably hot. When it got uncomfortably hot I would go to the pool and go to walk in. The temperature difference between being out of the pool and being in the pool made the pool seem much colder than it really was. It made it difficult to enter as it felt too cold and uncomfortable. I had to decide whether I want to be hot and uncomfortable or cold and uncomfortable. I knew that staying out of the pool would get hotter and hotter as the day went on, and more and more uncomfortable, yet also knew that once I was in the pool I would be fine, it was just taking the steps to get in the pool that was the challenge. In the end I decided 'sod it' and just jump in, and quickly got used to the water and feeling comfortable...

I told this story (not exactly the same, I tailored it to the client) to a client the other day. They were depressed, they seemed proud of how many Psychiatrists, Psychotherapists and Counsellors they had been to and that they had spent time in the priory and yet they were still depressed. They explained how they will 'always' be depressed so they 'have to get used to it' and that they were told they should see me just to 'talk it over'. They said they are uncomfortable with change and have tried CBT with no luck because they know what they should be doing and saying and they know that what they stop doing when they are depressed are the things that will stop them being depressed but they can't put all that into practice once the depression starts. They knew I had just got back from Dubai and so asked me how it was, what it was like (which is why that story came to mind)

The following week the client was much happier and cheerful (still has bits to work on) and she was using terms I used in my story to describe how she has been (like 'I decided 'sod it I'm just going to go for it' and feel uncomfortable mixing with people when I'm down because I know that will make me feel better')

The metaphor I used above is one I chose to use because it is a true event from my life that I can tell in conversation without it seeming like a metaphor, I'm just talking about my holiday experience. I find that the most important thing is to have a thread running through the stories you tell. So if you wanted trance you may talk about interests and as you talk about your own and the trance aspects of them (without mentioning trance if you want to be indirect), then you may talk about science (if you or they show an interest in that) and fascination with Newton under the apple tree and Einstein day dreaming travelling on a beam of light, you may then end up on the subject of holidays and so talk of trance aspects relating to holidays, etc...All these stories will make sense in context (EG; discussing interests, holidays etc.) with what is being discussed, also they are being discussed in a wider context of the overall discussion about the perceived problem so they will unconsciously make sense in relation to the problem. As each story has a same pattern in it (that of people entering trances spontaneously and effortlessly and positively etc...) the unconscious mind can spot this same repeated pattern as it is in each story. The same with hypnotic phenomena or patterns for resolving problems etc.

You can also be vaguer with patterns in stories especially when someone is in a trance, like stories of nature, seasons, animals, fairy tales, etc...

The unconscious is very good at working with patterns, so if you created a metaphor that laid down a pattern it can use that pattern in a different context (the problem context).

I worked with a French girl once that barely spoke any English and would have struggled to understand the words I was using if I used complex language patterns and may not know half the words, she wanted to quit smoking, she could speak some English so we could establish like and don't like. This was enough to start working with, the rest was images, holidays, demonstrating deep breathing in and filling lungs, not liked places, not liked images, shallow breathing and coughing and suggesting she should visit Arundel (a local countryside town), go to the top of a hill and breathe in some of that fresh air and wonder what it can mean in the context of being healthy. She stopped a few weeks later after doing all this and we had just enough language to get by...

Hypnosis, Trance Induction & Utilisation

One quick way to induce a trance is to have a person recall their problem (it is often likely to be trance inducing), like getting a smoker to recall smoking (or getting a craving), or a person in pain to focus on the pain (only this time in a non-attached way be focusing on its colour, shape, size, etc.), or a person that has OCD to discuss their OCD process, or someone with a spider phobia to recall the phobia, etc...

The higher the level of emotion the deeper the trance the person will naturally go into when they recall it.

You're always working with the trances you get, some people are just more responsive than others and so better hypnotic subjects.

Everybody is different, some people you can just look at them and say sleep and they will (if they know you do hypnosis and expect it to happen). Others would not respond in this way.

A good hypnotic subject is likely to be able to perform hypnotic phenomena and respond to therapy easily.

As Erickson has mentioned, in some cases he had to train people for some time to help them to be good hypnotic subjects. It is useful to know when someone is at that stage, so that you can move on to hypnotic therapy using different phenomena and so that you know they will be more responsive to what you say, whether this is when you first meet them or after you have trained them for some time. Generally though people don't need to be brilliant trance subjects to do good therapy, the therapist just needs to be able to utilise whatever the client brings to the therapy.

I naturally take fairly unnoticeable long slow breaths and people think I'm mucking around and holding my breath, this is often (not always) more pronounced when I enter trance. If I am hypnotising someone it is my responsibility to make the effort to match the clients breathing.

The trick to breathing quicker (but slow for the person your matching) is to just drag in and push out the air at a faster rate rather than do half of your normal breath then breath out (always leaving half your lungs full of stationary air because you never empty your lungs properly) as not emptying your lungs properly and filling them properly is bad for you. It is a bit like scuba diving and having to learn to control your breathing, then after a short while you can do it automatically.

If someone is breathing too fast or in a way that would be awkward then don't copy it exactly, you could do 3 of their breaths to one of yours (or any other comfortable option). And you could make emphasis to the out-breath and may be do your in-breath to 3 of theirs, then your out-breath to 4 of theirs.

There can be so many contexts when you want to notice as people enter mini trances so that they will be taking on what you are saying (assuming the trance includes you) or they could be in a trance to integrate what you have just taught (like doodling or staring into space) so you would want to give them a brief bit of time to finish. Or if you want to demonstrate and have as few problems as possible then someone very responsive is likely to carry out what you say best (which can also act as a convincer to the less engaged)

If you ask someone about the stages of their problem they have to enter trance to tell you. If you ask them about a leisure activity they enjoy they will enter trance. If you ask them what colour their front door is they will enter trance. Ask them how they will know when they are better and they will have to enter trance...

It would be difficult not to have them enter trance. Even if you sat doing nothing they will go inside to ask themselves what is going on, so they will have entered trance.

These are all small and can be built on and used for a bigger future trance, or any of these can be deepened as they appear.

When you ask someone 'have you ever been in a hypnotic trance before?' what you are doing is a double bind. This is because you have added the word before. If you ask have you ever been in a trance? They can say yes or no, if you ask 'before' it means before what? Before the one you are in? Before the one you are about to go into? So whether they answer yes or no they are accepting they will go into or are in a trance.

If they answer yes and it is a good experience then gathering information will quickly drop them into a trance again yet it will appear like you were just enquiring about that previous trance. If you want to still follow this line of questioning to induce trance when they have said no you can just explain what it will be like (using your hypnotic language skills)

Either way they are likely to enter a hypnotic state rapidly and be well on their way before they know what is happening.

My Experience of Stopping Using Scripts

I just wanted to share my experiences of stopping using scripts.

When I first trained everything was direct and all about using scripts. I even contacted every therapist in my area to learn from them, get their opinions and views on their success etc and all the feedback was to buy lots of scripts and when a client tells you what their problem is, use a script for that, find out which induction script they want and use that and use a script for ending the therapy. I had a collection of over 500 scripts! Imagine sitting with a client and trying to remember which script I should use!! I also felt it was wrong to just read in a monotonous voice from a sheet of paper and get paid for it and claim I know what I was doing. They could buy a book of scripts, choose the ones that suit them best, talk to a tape machine and do it themselves for much cheaper.

When I found out about Ericksonian Hypnosis I realised what Stephen was doing and Richard Bandler and that it wasn't that they had memorised inductions and therapy scripts and were reciting them, but that they were tailoring the therapy to the client.

I attended a two day course on Ericksonian Hypnosis and on the course we had to sit opposite someone and (like catchphrase) 'say what you see'. This was fine and I was comfortable with this in the safety of a course where at least I know I could do hypnosis, there were beginners that couldn't. I had also by this point started 'ad-libbing' self-help tracks because I couldn't find tracks or scripts for what I wanted to explore. I had also listened by this point to many of Stephen Brooks' Audio courses and seen numerous videos and so had a greater grasp of language patterns, tonality, etc...I still used scripts with client because I thought I would not know what to say.

After the course I met up with a friend that was willing to be a guinea pig, I said confidently that I can now do hypnosis without a script. I decided I would do a leisure induction with him and utilise his interests and times his mind has naturally wandered, and utilise on-going behaviours that I can observe.

I asked him 'in an ideal world where you could do anything, what would you do that would make your mind wander, that would make you lose track of time and really enjoy yourself?'

His response was 'I would go back to Thunder Mountain (apparently some water-ride in a water park in America?)'

I thought well I said I would use anything...so I did, and he said it was the deepest trance he had ever been in and we got numerous hypnotic phenomenon and great success.

I was nervous when he didn't say a nice warm beach or something like all the course participants had said, but I am glad, I have never looked back and now can't imagine using a script.

The thing I learnt is you can't be wrong because you are given your script moment by moment by paying attention. And if you expect them to go into a trance and so let your voice and breathing guide them it doesn't matter if you don't yet know all of the language patterns. You learn best by being uncertain at first rather than knowing it all then deciding to try it out.

There have been a few occasions where I have worked with people that need to know the side effects of everything. You talk to them and they tell you all about all the different tablets they take and how they always get most of the side effects. With these people on many occasions I have got them to be agreeing that when they receive treatments they have the side effects. I then give them side effects for the treatment they receive from me. These side effects are obviously positive though.

I do this when working with some people with Obsessive Compulsive Disorder also. I will give them a daily treatment plan that sounds specific but isn't, like between 1830 & 1945 you will have fun with your children, the plan gets followed obsessively, I have symptoms created of what happens if the plan isn't followed (positive of course) that gets the person trapped in a double bind. Doing the re-framing and getting agreement initially is the trick, once they are willing to follow the plan they also tie themselves into following the consequences of not following the plan...

Describing Your Own Experience to Induce a Trance

You know one of my interests is going on walks through the nearby woods. I'll spend hours just **wandering along** in my own little world...**feeling** the breeze on my skin...I...**begin to notice** the sound of each footstep...time seems to just... **slow right down**...and I seem to be able to ...**notice the smoothness** of the movement of my **breathing**, of each regular step, of individual sounds from the birds, the rustling of the leaves...**noticing** the shimmering rays of light...the **warmth** of the sun on my face...and as I continue walking I...**notice** how the **breathing begins to relax and deepen all by itself**...often I find my...**muscles relaxing**...around my shoulders, arms, neck and face...and before long it already seems like time to go home...

I find when I talk hypnotically about an interest I have the client often finds it a familiar experience and so gets guided indirectly by listening to my description. I did this for one person (a hypnotherapist) where I challenged myself to see if I could hypnotise a hypnotist without them noticing. Part of what I did was said 'you know I've always wanted to drive down America, see how things change on a journey through the States' I went into detail about this imaginary journey in conversation and he was in a trance in no time at all.

Regarding therapist entering trance as well as client, it is best when it happens. The difference is in the trance. The client enters a trance focused internally and the therapist (at least in my case) goes into a trance focused intensely on the client, paying full attention to the client. So the therapists trance is an externally focused trance, the clients internally focused.

Sometimes the therapist may not know all they need to know or they may not have time but want to get as much done as possible or they may have been presented with a number of issues and only worked on the one they could make change the fastest.

Nominalisations can be used to aid the client's unconscious to begin to spread change to other areas. I have recently posted a video on a site of mine where I work with a woman that over the phone said she wanted to quit smoking then came to me and said she wanted to lose weight and stop drinking cola and quit smoking. I asked which one of these was most important to her. Quitting drinking cola was what she expected to find hardest and was most important to her. I helped her with this issue whilst dropping in nominalisations and non-specific ideas for change in other areas to also occur. So far (three months later) she has lost about a stone and a half, cut down on smoking and had no problem stopping drinking cola with no side effects. She wants to now work exclusively on smoking in a follow up session. My aim was to promote a way for her unconscious mind to have permission and an understanding to spread change. Asking things like 'You can be curious to discover what other changes occur' A sentence with no specific meaning other than the one the listener places on it and it doesn't give any direction or content as to what is expected other than change. Given in a context where all change that is happening is positive the expected change is also likely to be positive.

In one session I couldn't cover all three issues but could indirectly begin to get movement on the issues I appear not to be working on.

Arm Levitation and Catalepsy

Just lifting an arm in an ambiguous way would induce catalepsy without asking for it (the movement would imply it).

Saying 'In a moment I'm going to lift your arm and I'm not going to tell you to put it down' Implies you are going to want the arm to be cataleptic but doesn't say this.

Telling a story about being in a cinema and your hand stopping in the air as something interesting happens on the screen implies catalepsy.

Saying 'and when I lift your arm you don't have to move it up, or down or left or right or in any other direction, you can just enjoy the relaxation' (implies it will stay still)

Talking about animals that lie in wait for hours on end without moving implies catalepsy.

The stories or metaphors above would be useful for seeding in advance, giving time for it to sink in then when you lift the arm the unconscious recognises the pattern and activates what was seeded earlier.

Another way could be to lift the arm so gently they client doesn't know if you are holding it or not so it gets confused and stays where it is.

Time levitation instructions/suggestions/commands with the clients in breaths...

It is a good relatively easy form or ratification keeping the arm in catalepsy when the person awakes from trance. I think it all depends on the person and the situation and what you want to achieve with it...

That is catalepsy, catalepsy is happening all the time somewhere in your body (like the neck staying in position to keep your head still). Catalepsy is not rigid like an iron bar (although this is often used as a metaphor) it is more like a waxy immobility that is comfortable, it is difficult to describe but your description is correct.

I have done full body catalepsy in un-hypnotised people by having them stand and then tapped on their shoulders in different directions causing confusion (like the tapping on the arm) and on the upper body and catalepsy sets in.

At the same time it makes the areas reduce in ability to feel sensation, they also stick where they are placed (if you lift a cataleptic leg it will stay where you let go of it, for example). It is good for initiating pain control or for operations.

In catalepsy there is no muscular forcefulness/tension, for example if the eyes are cataleptic it isn't like they are being held shut like when you tightly shut them, it is more like they just don't work.

I worked with one person that needed to believe he had been in a trance and I was videoing the session and I asked what would make him believe, he said if he could see on the video that he was in a trance, so I had him do catalepsy for the whole hour...He was convinced because he knew this was impossible normally, his arm would have wavered or lowered.

Other times I see that someone's arm is getting tired etc so I will suggest faster lowering, sometimes slower lowering, sometimes if they believe 'I have the power' and I need to be a bit more direct I wait until the arm is halfway down then I push it down to their leg as I link it to something internal almost like a shock/surprise induction being done with the person already in trance.

Other times I use it as a metaphor for something so it could be that I can lower the arm, then they can lower the arm (like lowering a resistance etc.) (with the arm placed between me and them)

Hallucinations

Suggestions can still be given indirectly, or priming/seeding can be done indirectly etc...

If someone doesn't see what you want them to see then you can reply with something like 'that's right, you really don't see it, and I wonder what else is there that you really can't see...'

The other way round it is to presuppose what you want them to hallucinate without saying it, like asking 'what breed do you think my dog is?' Whilst slowly gazing down towards the floor where you want them to hallucinate the dog. Or what do you think of my new picture? Whilst looking at a blank wall. If they say they don't know they can't see it cause confusion by implying not seeing it means being deeper in trance, and praise them for their ability to go so deep, and then deepen their trance etc...

If it was auditory hallucination you can mention how you can hear music in the background and ask if it is a piece they are familiar with, and then ask them to really focus on that music.

With hallucinations in most positive hallucinations is a negative element and vice versa, for example, if you hallucinate a chair in front of you; you have to hallucinate out parts of the background, if you don't see a chair that is there you have to hallucinate in parts of the background.

I once decided to do an experiment involving creating artificial auras. I hypnotised myself to see different colours around people for different modes. My logic was that there is so much information to take in (non-verbal signals, verbal cues, words, etc.) that I thought my unconscious is probably noticing all of this stuff, can it just process it for me and give me a cue that I can notice that sums up the information. What I thought was the best thing for this would be to see auras that I can observe changing and can work with (for example if I wanted a specific depth of trance, once the client is there the aura would be a dark blue, I this gets lighter they are coming up so I need to acknowledge then deepen, if it gets tinges of red there is some anxiety so I need to acknowledge this and deal with it etc.)

Over time the auras faded and I just started saying what came to mind whether it made sense or not.

I used a similar thing when I first started out doing hypnosis when I was about 14. I was envious of people with synaesthesia I thought 'if only I could see sounds, how useful would that be for playing man-hunt in the woods at night'. So this is what I did. I made it so that sounds would make light and so if someone stepped on a twig for example I would see a flash and know where they were. It is easier to judge where something comes from when you see it rather than when you just hear it. This is something that frustratingly I've not been so capable of as I've grown up.

Surprise and Confusion

Surprise or confusion can be used indirectly, just telling a joke can surprise. A handshake being slightly different is barely noticeable or paid attention to but it causes confusion and a trance, moving your head as you talk to the client (or looking into a different eye for conscious/unconscious etc.) causes some confusion as two messages are being conveyed with different meaning, one of them to the unconscious to go into a trance, overloading the client with information causes confusion, like asking them to do something then before they have time to get it done ask for more and more until they need to take on some of the tasks unconsciously. making purposeful mistakes can cause confusion like saying I'm going to reach over and lift up your left hand' and lifting the right one, or saying different to what you are doing 'I can lift your hand up, down left right etc (doing as you are saying) then after a few rounds of being congruent move the arm different to what you are saying (up, move hand down, down move hand up, left, move hand right etc.

When I tell confusing stories changing terms with one meaning into characters people seem to think it is a challenge to do. The trick is to turn each term into a character then just tell a story, it isn't confusing to me saying it (unless I rattle it off too fast) because it is just a story and characters.

Left/right confusion etc. is all something you go into a trance to do. I normally do small doses not long reams although I don't have a great problem with this. It has to be right for the client and context.

Client says left hand feels heavier than right, I might say so your left hand resting right there is heavier than the right land left right here is that right? Etc...

Compound Suggestions

Compound suggestions can overlap. Generally it is a truism followed by a suggestion; this can be from observable to non-observable, out of trance to in trance, etc...

For example:

You can sit there, and read this writing

You can read this writing, and let thoughts come to mind

Those thoughts can come to mind, and some can be of pleasurable experiences

You can be aware of those pleasurable experiences, and become more absorbed and relaxed

One thing I did when initially learning this and all of the other language patterns and structures etc. was to listen to conversations (in real life and on TV etc.) and look out for specific patterns.

In work lots of times people would say things like 'Your shift doesn't finish for another hour, does it? Can you go get the paperwork up to date' Implying because the shift doesn't finish the person can do the paperwork although there is no real link between the two.

In ordinary conversation people don't often work from observable to non-observable, or from not in trance to in trance. (Some good communicators do) Normally it is just truism-suggestion, sometimes they can be linked but most people don't realise they are doing it so just use single sentences.

Another one could be

'You know where Johnny is? Can you call him for tea'

In sales

'Look at this phone; it meets all of your needs'

'You look like someone that likes making good decisions; this is the TV for you'

'You want the Big Mac Meal, and you're going large with that' (Question said as a statement)

On TV

'The question is shown on the screen; phone in if you know the answer'

'It's the end of the show; enter this competition to win £5000'

Contingent Suggestions

Some examples of contingent suggestions you may hear in everyday situations:

You don't have to brush your teeth until you're about to go to bed

When you go to the shop remember to get some milk

Wash your hands before you eat dinner

I'll read you a story when you're in bed

You can have chocolate fudge cake after you have finished your dinner

Contingent suggestions make one part of a sentence contingent on the other. The way to word them is to have the contingent part an unconscious process. If it is unconscious the client can't say 'no' when the behaviour it is linked to is true and happening

'As you blink in that special way you can become more absorbed'

'As you breathe out you can relax deeper'

'As you look at me, you can also be aware of certain thoughts that come to mind...as you become aware of those thoughts you can wonder what is happening in those hands...as you wonder what is happening in those hands you can notice that one hand feels different from the other...'

All pacing and leading and all starting with a truism. The contingent parts are all out of conscious control. Becoming absorbed, relaxing deeper, having thoughts, wondering what is happening in the hands, hands feeling different from each other. Nominalisations obviously help here with the leading parts.

When are People in a Trance Naturally?

I would say when people lose car keys that are right under their nose they are in a trance state. In the same way that when you get catalepsy in a cinema you are in a trance, when you forget a name at a party that you know you know and the harder you try to recall the name the more elusive it becomes, you are in a trance state.

I think it all comes down to how you are defining trance. If you get into a state of uncontrollable laughter you are in a trance state, same with problems like depression, smoking, anxiety, etc. they all involve going into a trance and at the time you are in that state you see the world through that trance. Change trances and you see the world differently.

With the key example assuming they are in view and somewhere you have looked and not seen them often people find it is when they need the keys, they focus all their attention on 'where are the keys', then when they can't find them focus on how they can't find them and begin to narrow their attention on the issue of keys missing and as heightened emotion also induces deep trances (like phobias, fetishes etc.) they are now getting emotional (stressed, anxious etc.) because they can't find the keys so the trance gets deeper and more powerful and it cycles round as a self-fulfilling prophecy. The answer (as with the name example etc.) is to stop, and think about something completely different to break the cycle.

One thing that gave me confidence at inducing a trance was seeing that everyone was going in and out of trances all by themselves all the time (also see Rossi; The 20 minute break - a book about Ultradian Rhythms). Leisure activities induce trance, reading, listening to music, daydreaming. Most trances people go into are self-induced and most people wouldn't notice that someone is in a trance because they wouldn't be looking for it. Driving involves trance (sometimes deeper than other times) you have to do many tasks simultaneously without thought, same as tying shoe laces, doodling, all automatic behaviours involve mini trances (handshakes, etc.) A hypnotherapist can interrupt these trances and extend them and become a focal part of them to take control of the trance.

Binds, Double Binds & Implication

If you use things like a double bind you presuppose one direction whilst they think they are always making the choices which you are then responding to.

I remember watching Erickson say to a client 'look at my dog, what breed do you think it is?' The client wasn't asked if a dog was there only the breed.

If you are stating truisms people are not necessarily going to notice you are using a technique, and they can't really find holes in it. Also if you ask questions with implied responses but not actually asking for verbal responses or questions they would seem stupid to answer no to (like 'so your name's Steph?', or clarifying age etc...) it can just seem like you are clarifying rather than trying some technique.

Implied responses could be

'So you're sitting in that chair, and you can notice me and hear my voice and you don't expect to go into a trance yet' (Four implied agreements that don't ask for a response so are unlikely to face resistance etc...)

Post Hypnotic Suggestions (PHS)

Whenever anyone carries out a post hypnotic suggestion they go back into a trance like they were in when the suggestion was created to carry out the suggestion. If this trance is interrupted before finishing the PHS then you can expand on it and utilise it.

If someone has constant pain and it is ok to remove or alter the pain then you may want a client to hallucinate numbness or a different sensation for a long period of time (perhaps with conditions that if the signal is required it will come through)

You may not want to tell them it is only an hallucination and you may want them to be stuck in the hallucination to the extent that the pain control lasts a while. And that a trigger like opening the eyes in the morning could be used as a PHS for the hallucination to begin each morning.

At the end of therapy you would end everything you don't want them leaving therapy with so that they are completely reoriented back to 'reality' before going home. You may say 'you can wake up totally and completely' or 'wake up all over'

Nominalisations

Another area with therapeutic nominalisations is building your own context through the links between the nominalisations. If Development was used with talk of business the meaning of development to the listener is more likely to be in the context of business and it could be good or bad. If development was used in the context of 'what is happening now' then it is more likely to bring up meaning in this context.

The context the nominalisation is given in effects the meaning of the nominalisation.

'New developments are happening in the business, there will be organisational changes taking place'

'New developments are happening inside your mind, and you can wonder how those organisational changes and improvements will take place'

Fractionation

Fractionation is where you put someone in a trance and out of trance repeatedly. Each new trance induction deepens the trance. Erickson noticed each time he hypnotised patients they would go deeper than the previous time. He would spend weeks (sometimes many months) hypnotising clients. Then he wondered if he did the same number of inductions in a session rather than over many sessions would it have the same effect and he found out it did. To make it more effective it is useful to not bring the person 'out' between inductions. Just distract them (if they have their eyes open) or ask them to open their eyes (and not ask them to wake up etc.) and talk and then do another induction (all can be indirect or it could be overt and asking them to 'close their eyes')

Inducing Trance with Music

You can induce trance with music. Many cultures have used music and no words to induce altered states (trance states). Many tribal cultures use flickering lights of the flames of a fire and drum beats to induce the trance state. Sometimes this beating can be fast, other times it could be slow. The trance states are different depending on the speed of the beat but it can all be used. I often use drum beats and other sounds to induce trance on my mp3 tracks and use binaural beats (different frequency of beats to each ear to create an illusion of a beat out of the difference between the two beats that often leads to the frequency of the brain matching these beats)

This all happens in music, especially modern music that can be listened to in stereo (or 5.1 Surround Sound) that allows musicians to create tracks that allow for deeper absorption from the listener. So any music can probably be used to induce a trance. I've known a number of teenagers I've worked with that would use Eminem or a gangsta rap group...If anything (especially with this age group) it shows respect to them that what they like and are into isn't being dismissed. I feel the times I've used these musicians/types of music that if I dismissed it and used something else it may have appeared disrespectful. I appear to show a keen interest if I ask for more detail and use it. You don't even have to know much about what you are discussing because you can be vague with your language and use nominalisations so it sounds meaningful to them.

Richard Bandlers Neurosonics hypnosis tracks use Blues and jazz to help induce a trance...I have used marching as a trance induction (ex-soldier), I have used dance music (trance music) and even rollercoaster's, through to diving, hang gliding, etc...Almost anything can be used...

Every day Trance Phenomena

What is commonly thought of as deep trance phenomena can occur in a lighter trance, like catalepsy occurring when watching a film in the cinema, or someone hallucinating that they thought they heard someone say their name, or hallucinating that they saw something out of the corner of their eye after watching a scary film (obviously these examples of naturally occurring times and so aren't as dramatic as when under hypnosis, but they are still occurring). I have had arm catalepsy in people in seconds while they are in a light trance. Noticing multiple deep trance phenomena or signs is more likely to mean deep trance is present than just noticing one (same as trance indicators like REM, fluttering eye lids, etc.).

Rapid Inductions

To do them the main thing is confidence. It is about interrupting a pattern or causing confusion. I rarely use rapid inductions. I rarely in private practice use anything that resembles a formal induction. At best I often ask people to close their eyes and I begin to talk to them and observe them closely, slowing down my breathing and speech, lowering my voice and tonality, and I mention talking to the unconscious part of them and just do it.

Really the inductions I do don't exist. For example if I am treating a phobia I just say something like 'OK just close your eyes a minute and we'll try something' then I go into the technique and make sure that all of my non-verbal behaviour is implying trance induction and deepening while they focus on the words and following any instructions (which is also trance inducing).

Watching Bandler, or some of Milton Erickson's footage can help. Also reading 'The deep trance training manual' by Igor Ledochowski and 'Training Trances' by Overdurf & Silverthorn is useful. These have info about rapid inductions and handshake inductions.

I do handshake inductions similar to how Erickson did them as they are less dramatic, or I will tell someone that 'in a moment I'm going to reach over and lift up your right arm...and I'm not going to tell you to put it down (implying levitation)...any faster than...your unconscious (embedded command)...allows you to go deeply and comfortably into a trance while the conscious part of you can drift off and think of pleasant memories or hopes and dreams...etc...' Then I reach over lift the arm very gently so I am hardly touching it and once it is levitating I say 'that's it' or 'that's right', as it starts lowering I suggest it can take its time etc... The person will already be in a trance, and as the arm lowers they will go deeper.

Or I do the fingers coming together induction. This induction is rapid and the client can then recreate it (if this is suggested) as a self-hypnosis induction in the future...

Observation Skills

When I first learnt about congruency between conscious and unconscious messages I wanted to know how I could practice this and refine it as a skill.

The best way I have found is to watch people, watch them in pubs, clubs, restaurants, anywhere where you get to observe people interacting. By doing this you can listen to conversations at the same time as objectively watching non-verbal behaviour. Another place to watch this is on reality TV shows like Big Brother and on programmes like 'the Jeremy Kyle show'. I used to record one of these shows a week and watch interactions and see what I could figure out about people based on mismatching communication. With programmes like Big Brother you can test your ideas about your observations over a period of time.

You can watch people talking and look for patterns. Doing this you don't get to ask the questions but you can pay your full attention because you aren't involved. Anyone that has knowledge of magic and watches a magician knows that if the magician is captivating enough you miss what they do even though you know it happened right under your nose. This is the same when starting out doing therapy, you know lots of stuff but miss it when you are in a real situation because you have too much to take in.

As you watch people you may work by initially just getting a sense of something or you may actively look for patterns that you could tell someone else (like change in facial colour, change in lips, body posture, eye contact, etc.)

The best way to learn to recognise minimal cues is to focus on one at a time while you learn.

What you do with the observations depends on what you are observing for (it could be to look for congruence, or it could be for a specific state, etc.) If it is for a state then you can suggest back the minimal cues, so if you wanted to induce a deep trance comment on the minimal cues (overtly or indirectly) each time you see a trance based minimal cue. You could link it to going deeper for example by saying 'as you continue to blink in that special way you can drift deeper.' Or 'That's Right' (said on each blink or sign of ideo-motor movement etc.)

The easiest way of noticing minimal cues is to be in a trance, letting your unconscious notice for you.

To switch the trance focus (from internal to external or external to internal) you can start by matching the experience then guiding it to where it is wanted.

'You can be aware of the ticking clock, of the traffic outside, of the sound of my breathing AND you can notice what those hands feel like resting on your lap WHILE you wonder what will happen next...and BEFORE you discover what will happen next you can notice which hand feels the heaviest and wonder which one will lift...' (Getting more internal)

To do this the other way reverse the process and match on-going internal experience then you can ask them to remain in this state while they open their eyes and pay their full attention honestly and completely (a statement they should take literally) to ... (whatever the external thing is - reading, practicing an instrument etc.)

With leisure activities you can have an external focus activity and guide it internally (even by saying 'I sense you can feel some of that now').

Analogue Marking

With the issue of analogue marking or embedded commands, I use it all the time throughout whole sessions. It is like allowing your communication to be multi layered. The conscious mind will be phasing in and out.

I think it is important to be continually allowing communication to both the conscious part and the unconscious part. Analogue marking allows the conscious part to listen into an apparent conversation whilst the unconscious part is aware of the marked out sections.

Because the unconscious part notices these sections and the conscious part doesn't the conscious and unconscious receive two different messages. I often do this telling stories/metaphors etc. that the conscious mind just listens to while the unconscious part responds to the patterns in what I am saying and also to any sections that get marked out (a form of analogue marking is embedded commands/suggestions).

The 'my friend john' technique is a good example of this used in trance induction. It also happens in everyday life, you get people that say 'I told him...I'm really annoyed at the lack of respect you show me...' As you hear someone talking directly at you like this it can feel like it is aimed at you, it creates feelings in your body as it affects you on a deeper level even though logically and consciously you know they are not talking about you.

If ideas and suggestions are given indirectly (via analogue marking for example) then the conscious mind is highly unlikely to notice so it will only be received unconsciously, if the suggestions were given directly then the conscious mind may become aware and may in the future sabotage the work because it remembers bits and pieces.

When I want to educate someone in therapy and feel that they probably don't see that they don't know what they don't know I do it indirectly. Often by telling them I'm not going to tell them. because I am there as a therapist and they have come to me for help there is a high chance that they expect me to know what I'm talking about so if I don't give them reason to challenge me then often what I say gets accepted.

As an example in smoking, some people think they know the risks of smoking but don't really, they only know the common few things that get plastered over the media. I want them to have an understanding of some of the other issues but I don't want to lecture them or to have them defend why it won't be them etc (I don't always feel this is necessary, it is client dependant).

I will often say 'I know you know all the effects of smoking, so I don't need to tell you that 50% of all smokers die of a smoking related illness'...then I tell them what I said I wasn't going to tell them.

Rapport

Over the years I have found that many people misunderstand what they should be doing when building rapport and matching or mirroring body language etc. People seem to think they should be copying but this really annoys people. The idea is to join the person in their reality respectfully. Look for patterns. If they do a specific gesture when saying or talking about a specific thing (like making a brushing motion when talking about getting rid of a habit) then if you talk about the same you can make the same motion. You don't just do it because they did.

I match breathing and often match heart rate with the movement of my head or a finger or my foot, often match blinking (either with my blinking or with a finger etc.), I match general body posture and certain words and phrases used. I don't match things that would be unnatural for me (I had a client that had a bad arm and she kept it in an awkward position. I could copy it but it would have been uncomfortable and would have looked like I was taking the piss).

If you are matching how someone is sitting and they change positions, don't immediately change how you are sitting, wait until a time when it would be natural for you to decide to change positions.

The idea is to respond in the same way they would, not to copy them. I pace then test then lead people into trance without placing any importance on the words I am saying just by using breathing, body posture, heart rate, blinking etc. so that when they are responding to me I gradually put myself in trance and they follow.

Therapy in Action – Performance Enhancement: A Conversational Hypnosis Demonstration

This is a transcripts of a therapy session that includes my analysis to show what is going on to create change and to explain a little of what I am doing.

In this session I am helping the client to improve their artistic abilities. I have looked at research into creating savant abilities in people and thought it may be possible to do this with hypnosis.

The session was an hour long and the only session required to help this client improve their artistic abilities and to still (about 5 years later at the time of writing this) maintain and develop that improvement further as time has gone on.

I use a 'D' for when I am speaking and a 'C' for the client. The analysis of what I am doing is 'cut in' to the session where I have felt it is useful to note techniques or language patterns I am using. Hopefully this session will give a good overview of how the techniques and learnings from this book can be applied during therapy helping to bridge the gap between theoretical learning and practical application of hypnotic and therapeutic skills.

Art Improvement Session

I ask the client to draw a picture of a horse in one minute and at the end will ask them to do the same again.

While this session was taking place I am matching the clients arm positions, breathing, leg positions and leaning in at about the same angle the client is leaning back. This all helps with rapport and building a deeper connection with the client and their model of reality.

Initiating Trance

D: (Looking down from the client) When was the last time you (looking up at the client) went *into a trance*

C: Mmm...I don't know...probably in mmm, I went into a trance...well...probably last week...Tuesday...Tuesday this week because I was pruning my bonsai tree

D: When was the last time you (looking at the client in the same way I did above) went *into a deep trance*... (looking down) do you remember (looking back up at the client) *what one is like*...

C: Yep

D: *You do*

C: Vaguely

How I often start out inducing a hypnotic trance is by getting the client to recall a previous trance experience. If a client says they have never been in a trance before then I ask them what they think it will be like or what they expect it to be like; or I ask them about everyday trance states like leisure activities.

All the highlighted words are embedded commands or suggestions. They are parts of the communication I am adding extra emphasis to by using a more hypnotic voice, defocusing my eyes and relaxing my facial muscles all to imply trance through modelling what I am expecting.

I imply that the client has got experiences of entering a trance and deep trance by using the term 'when'; I don't ask 'have you ever been in a trance?' In some cases I may ask 'have you' if I strongly suspect they may not understand or realise that they have been in a trance before. If I do this I would use the term 'before' at the end of the sentence as 'before' implies either before the one they are in now or the one they will be going into.

I very often use feeding back as a way of embedding suggestions or commands; for example when the client says 'yep' and I respond with the command 'you do'. I will often turn questions to statements with the tone of my voice. By feeding back what was just said you also begin to form a 'yes set' this is where you get yes responses that help to build rapport; displays understanding and makes it more difficult for the client to respond negatively to the work you do. If you get someone to say three or four yes's then give a suggestion you want them to follow they are more likely to do it than if you just gave them the suggestion on its own.

This client showed signs of not wanting to commit just yet by responding to my feeding back with the response 'vaguely' rather than responding by saying 'yes'. He had just answered my question by saying that he can remember what a deep trance is like but almost immediately became indecisive.

Seeding/Priming Using Metaphor

Sowing the seeds for amnesia and entering a trance when drawing

D: (starting to tell a true story to seed what is to come in the session, as I do I look down) You know when I was a kid I used to like watching…I can't remember the name of the programme now…that Rolfe Harris (turn to face the client) **drawing**… (Keep facing the client but move my head position) kids programme that was on CITV…

C: Rolfe's Cartoon Club

D: Yeah something like that (I look down again) he used to have that like (drawing a circle with my arm and finger) (looking back at the client) **circular desk thing**

C: (Doing an impression of Rolfe Harris)

D: (looking down again) And Mmm I (looking back at the client) got the book of that (Looking down again) I don't know how and I don't know what happened to it…it's not like me to lose a book…mmm…but I (looking back at the client) **always wanted to draw** the Rolfaroo (drawing in the air again)

C: Ahh yeah

I now use a true story to begin to plant the idea of automatic drawing. I also want the client to have amnesia for much of the session so that later they don't try to review everything that I was doing. This specific client has considerable experience with using hypnosis so I want to make sure they don't analyse the session too much when they come out of a trance. I also want to tell a story that lays down the pattern of the session we are in now where I want the client to start off more conscious, then the unconscious can take over more. I also have chosen to use a story based on being a child of about eight years old because that way the client will begin to make associations unconsciously with when they were eight and this will cause a slight regression and will change the clients beliefs and opinions of what is possible; as an eight year old almost anything is possible.

I am still continuing to embed suggestions and ideas. While I am telling the story I am drawing in the air (like automatic drawing taking place) and putting emphasis on the words I am saying while I do this so that the client knows that these parts are important.

D: And in the book (gesturing where the book is) it tells you how to draw it, but I could never draw it, I could never get it to look like the one Rolfe Harris (looking at client) **draws** (drawing in the air again)...(looking away) and I was 8 or something...didn't matter how hard I tried to draw it, it never looked like the ones Rolfe

Harris draws...then one day I was in the lounge trying to **draw it** (drawing in the air)...kept failing, well not failing, I kept drawing it but not doing a good job...so I sort of gave up and still had the paper and still had the pen on me, and was watching something else, and then I thought 'yeah **lets draw it again**' (mimic going to draw) (a car alarm goes off)...then I got distracted, by the TV...I can't remember what I was watching now but it distracted me and I suddenly thought 'cool I'll watch that'...**didn't even think**...It was one of those things where you know if you put an action movie on like Die Hard or something, and not a lot's happening so you're happy to read or do whatever, and then all of a sudden you hear some action happen and so you look up at the screen and think 'cool' and mmm so I looked up at the screen but my **hand kept drawing**...and I looked back down...and...like the cinema thing...you can be eating popcorn and something exciting happens and all of a sudden your hand freezes in front of your mouth because you're more concerned about (gesturing in front of me and looking in front of me – congruent – and gesturing holding a piece of popcorn in front of my mouth) what is going on on the TV and all of a sudden that finishes and you carry on putting the food in your mouth (gesture doing this)...as if...**you know**...the interruption never happened...and but what happened was my hand didn't stop, so when I was interrupted and I looked up, my **hand continued to draw**...and I looked down at a perfect Rolferoo and I was really proud except that it surprised me because I couldn't do it...(simulating trying to draw) never mind how hard I tried to draw it I couldn't draw it, and it took me by surprise because I did it without looking so I didn't know where anything was, so I shouldn't have been able to get the eye in the right place and I didn't know where the paper was...and people touch type and they can't see where all the keys are...they can get their fingers to the right keys...they can't see it at all...they can do it with their eyes closed and you know people play the piano the same...and people play the piano blind, like Stevie Wonder...so obviously it is possible for your body to know things that you don't know...like my granddad plays the piano without looking at the keyboard...and it is just one of those things...yes so Granddad can play the piano and not look at the keys...mmm could probably play in his sleep...and it's always fascinated me, I can now...I've never learnt to touch type, but I can now type to some extent without looking at the keys, as long as I don't consciously make the effort not to look at the keys, if I just glance up at the screen I can keep my fingers going (acting out typing without looking at my fingers) ... If I consciously think 'Oh I've been typing without looking at my hands' all of a sudden I'm pressing all the wrong buttons trying to type...

As the story continues I talk about how trying hard can lead to failure, or at least not getting the success you want. This is important because trying is a conscious process. And generally the harder you try the harder something seems to become. In the same way that if you try to fall asleep you struggle and stay awake, and if you try to stay awake you fall asleep. What I want is for the client to just do, not try.

When a car alarm went off outside I decided to incorporate this into what I was saying so I spoke at that point of the distraction that gets your attention for a little while. I then continued to parallel the alarm by saying about an action film like Die Hard and hearing something happening that gets your attention. While I told this piece I suggested amnesia and also suggested it being a positive distraction by interpreting the distraction in my story as 'cool'. I then talk about the hand continuing to draw while I am distracted partly because I want this process to happen in the client when he draws but also because if the alarm keeps going off or goes off again I want the client to be still responding on an unconscious level regardless of external distractions.

I create frustration in the story so that the client would become more focused on what I was saying. I spoke about how my hand continued moving then I ended that piece at an 'and…' before changing to talk about being in a cinema. What this does is a part of the client stays focused on looking for the completion of the pattern/story; and while that is happening the client enters a slightly deeper trance giving me the opportunity while they are partially listening and partially focused on waiting for the end of the story to continue to embed the idea of amnesia that I have been embedding already and also metaphorically commenting that the car alarm has stopped now.

I then carry on telling the story to convey the message that while the conscious mind is occupied with one thing the unconscious mind can work on something else. I then give different examples of being able to do things without conscious attention. Each example is true and undeniable which helps to convey the message that what we are doing is also possible. I use these stories to convey the message of not paying conscious attention, being pleasantly surprised by what the unconscious has done, which also contains the message that the conscious won't be a part of the process.

Towards the end of the story I start to separate conscious and unconscious by talking leaning slightly in one direction when talking about the unconscious mind and the other direction when talking about the conscious mind. I say '…it is possible for your body to know things that you don't know'. As I say this I spatially mark out; with my leaning; the body knowing and the mind not knowing.

I continue to seed the idea of doing things in a trance by talking about how my Granddad could probably play the piano in his sleep. I also suggest that conscious interference can stop this working just by having that conscious awareness; what is wanted is to just be in the moment and let it happen without thinking about it.

C: That sucks yeah, but you're right you get similar sort of feelings like when you're trying to draw or copy something and it doesn't do what it's supposed to do, I'd like to er, I don't know, say get a Disney video and be able to just put it on the side and just copy the video, that would be cool, I know people that can do that really really well and I can't understand why they can do it and I can't

D: Yes so it is possible obviously for your body to do things that you don't have an awareness of and obviously consciously you take in a certain amount of information, or at least you don't take it in, you take it in unconsciously and you're drip feed a certain little amount at a time and a very small amount of information, so you don't notice the fine detail, you don't really sort of, you just let the hand do what it is doing then it doesn't really matter what your conscious mind is thinking, what's on your conscious mind because your hand can just do it...any way that was just an aside...so you haven't been in a deep trance for a while?

C: No

In response to hearing my story the client agrees with what I have said and then states a goal that he would like; that would demonstrate success. He also says that he knows people that can do what he wants to do; which means he knows that it can be done. He also states that consciously he doesn't know how to do this; we are hoping to get him doing this without conscious awareness of how he does it; we just want him to do it.

My response to his statement was to focus on the unconscious process we are laying down and to link it back with him. He had spoken about 'other people' I wanted it to be him that has these abilities so I speak about his body doing things the conscious mind has no awareness of; again I do this by marking out the conscious and unconscious so that when I talk to different locations he will know I am communicating with him on a conscious level or an unconscious level. I then very clearly state what will be expected;

that he will notice fine detail and that what his conscious mind is doing has no relevance on his unconscious mind drawing as long as the conscious mind just lets it happen unhindered.

I use the term 'obviously' to convey the meaning that 'this is common knowledge, everyone knows this' this is to de-potentiate any resistance by using a term that is unlikely to be questioned. Generally people don't question something that they feel they should know and they often just let it go in because you are not asking for a response. When I work with smokers I often do similar; I will say 'obviously you know all the dangers of smoking, so I'm not going to tell you that...' then I go on to list a load of dangers without receiving resistance because I have framed the statement as just telling them what I am not going to tell them.

Induction

D: There was something I wanted to do and the question is do I do it now or do I not? But I might give it a go...

C: OK (looking confused)

I mention the induction I am going to do in this way as the client has been hypnotised by me before and I didn't want him to know what to expect. I wanted to begin to indirectly hypnotise him. The only way for the client to respond to what I said was for him to go inside his mind to wonder what I am thinking of doing and be confused because of not knowing what to expect. Most effective inductions have an element of confusion to disrupt the conscious mental set as the only way to respond to confusion is to try to find a way out; which will either be by following a clear suggestion by the Hypnotherapist or by going inside their own mind to escape the confusion.

D: Mmm...only because it's just something different for a change...

C: nice day for a change...

D: Yep I was just thinking exactly the same thing….right put your hand on my hand...OK...in a minute I'm going to tell you to push down, now I don't want this to surprise you...so I'm going to tell you in advance what's going to happen...

C: Alright

I mentioned 'something different for a change' so that the client would think of the sentence 'nice day for a change' which implies change happening. I also imply that because I am doing something different that will create the changes we will get in this session.

I tell the client what to do at the start of the induction I will be doing with him and even though I am about to do a very direct induction I still use indirect methods; I tell him I don't want the induction to surprise him and that because of this I will tell him in advance what will happen. I could have just told him directly 'when I do this you will go into a trance like this...etc' but he may have chosen to resist me if I did that so by framing it that I am telling him for his benefit so that he knows what to expect he is more likely to listen to hear what is going to happen rather than feeling that he is being told what to do and feeling manipulated. I then change direction so that he doesn't have too long to analyse what I have just said in case he picks it apart and realises what I am doing. I don't just change to any old subject, I change to the issue of trance, getting some agreement, creating some confusion and again getting the client to recall a previous trance so that when I start the induction he is already partially in the recalled trance and mildly confused making him more receptive and has agreed with me a few times enhancing his receptivity and making him more likely to continue to accept what I say.

D: And you don't mind **going into a deep trance** do you?

C: Nope

D: You really don't mind?

C: No

To make these sentences more powerful and less likely to be resisted I have used the 'Reverse Yes Set'. The reverse yes set is where you get a no but it is still in agreement with what you have said. By framing the questions negatively the client is less likely to feel the need to disagree. By asking 'You really don't mind?' The client is likely to go inside their mind wondering is there supposed to be a reason to mind? So this causes slight confusion for the client because why would they mind? While they are inside their mind searching previous trance experiences to make sense of the question and see if any previous deep trance experiences should lead them to a different answer they are putting themselves into a deeper trance just before I start the actual induction.

D: What's the deepest trance you've been into? Maybe the one where you lost track of time, although you knew exactly what the time was because you came out of the trance at the right time…Right what I'm going to do is tell you to look into my eyes…not yet…so what I'm going to do is tell you to push down on my hand (client starts pushing) in a minute…in a minute…(client laughing)…and…then when I say sleep (clicking my fingers)…sleep…and do it your own way…you can go relaxed, feet up, head down, your choice, you can slump, do whatever you want, however you want to go into it…OK…you happy with that.

C: Yep

I then immediately ask 'what's the deepest trance you've been into?' And then begin to describe a trance experience suggesting that the client can go into a deep trance where he stops paying attention to conscious references like keeping track of time, where he can just be in each moment; and that he can come fully back when he is supposed to. I do this to keep the client occupied with thoughts of being in a trance. He is already partially recalling a trance experience from the last questions I asked; now he will be beginning to recall a specific deep trance experience.

I then jump straight to describing what will happen with the current induction. All of this jumping around keeps some confusion there and keeps the client frustrated needing to become reliant on me for clues about what he is supposed to be doing. The client demonstrates that he is already becoming very receptive by responding literally to my description. He looks into my eyes when I mention it; and pushes down on my hand when I mention that.

D: Right push down on my hand...really hard...push, harder, keep pushing, harder, harder, keep pushing harder, harder, push down harder, harder, look into my eyes, keep pushing down, harder and harder, harder and harder

C: I can't

D: Keep pushing harder and harder, close your eyes and sleep (clicking my fingers as I say sleep)...

This induction is a very typical rapid induction. It is done very directly and forcefully. While I am doing this induction I am closely watching the client. I am watching his face and shoulders looking for changes that will signal that he has gone inside his mind because it is at that point that I want to give the command to go inside his mind and sleep. After about 30 seconds the client says; just audibly; 'I can't'. It is at this point that I want to offer him a firm suggestion to sleep. When he says he can't he has almost given up trying; he has become slightly confused as how can he push harder if he is pushing as hard as he can. When people are confused they will usually accept any firm suggestion so long as it doesn't go against their personal values. In this induction that is when I say just once more 'keep pushing harder and harder' just to add to the confusion; then I say firmly and directly 'close your eyes and sleep' at the same time as clicking my fingers (which will trigger the reorientation response which is the response that gets fired to alert us to a stimulus making someone focus on just one thing; this fires as we enter dreams which is what can give the sensation of falling); I also pull my hand out from under his which also causes shock creating a trance state. With the shock from the click and from the hand being removed and with the confusion and the direct command the client was very unlikely to not go into a deep trance. The client slumps back into the chair as if all his muscles have relaxed and his arm falls to his lap.

Trance Deepener

D: That's it, just allow yourself to go down deeper and deeper...that's it, deeper and deeper, just taking deep breaths, that's it......that's it...

I now deepen the trance very directly by telling the client to 'go down deeper and deeper' and saying 'that's it' on the clients out breaths. I also say 'just taking deep breaths' the reason for this was that by saying 'just' implies he is only going to take deep breaths not any other types of breaths.

D: And in a moment I'm going to lift up that hand (looking over at the clients hands), and when I do I'm not going to tell you to put it down, any faster...that's it...than your unconscious mind begins...that's it...to get a sense of what it's like to walk down a flight of stairs...

I continue to frustrate responses to help continue to deepen the trance by being ambiguous; not telling the client which hand I will lift; by saying in a moment not specifying exactly when. I tell the client indirectly that I want him to do an arm levitation by telling him that 'I'm not going to tell you to put it down'. I follow this up with 'any faster' which implies that it will lower by itself (which as the sentence continues links the lowering with the unconscious walking down stairs).

I continue to say 'that's it' on each out breath by the client as I am still deepening the trance so I want to use everything to do this.

D: And I'm not going to know where those stairs are...I'm not going to know whether they are in a building or whether they are going down to a garden...that's it...or down to a beach...only you know...that's it...where those steps are...

I emphasis me not knowing as this implies someone must know and if it isn't me it must be the client. I then tell him that he knows. I find that often it is best to indirectly imply something and lay down a pattern before saying something directly so that you are constantly communicating on two levels; one with implication and metaphor and patterns and the other being direct and able to be understood consciously. Even if I am talking to the unconscious mind I know the conscious mind is listening to some extent so it needs to have a message to follow that seems straight forward enough to not feel a need to analyse too much. Likewise when I am talking to the conscious mind I know the unconscious is listening.

Contingent Suggestions, Compound Suggestions & Nominalisations

D: And the conscious part of you that is *normally* (emphasised) at the front of your mind can just distract itself in some way as you go down those stairs and it can find its own way of becoming more and more distracted with each step it takes

Contingent suggestions are suggestions where two parts of a communication are linked with an action phrase like; before, during, after, while, as. The two parts don't have to genuinely be linked or relate to each other. In the sentence above I link the conscious mind becoming more distracted with walking down the stairs in the mind even though there is no real link other than me saying 'as'. I also include a compound suggestion. Compound suggestions have two parts of a communication linked with an 'and' or a pause. They imply that because the first part happens so will the second part. In the sentence above I use 'and' to link the going down the stairs with becoming more and more distracted. In effect I am saying the same thing twice in the same sentence using two different language patterns.

In this sentence there is also the implication or presupposition that the client will be walking down the stairs and that he will be consciously distracted. The first half of the sentence was saying the conscious part can distract itself, the second half was about 'how' this distraction will happen. The idea behind this is that the client is now more likely to focus on the 'how' rather than whether the distraction will or will not happen. There is also the implication that the client is in a different state and things that happen in this state are different to normal waking state by emphasising 'normally' implying this isn't 'normally'.

There are also many nominalisations just in this one sentence. Nominalisations are words with no fixed meaning. The listener makes up their own meaning when they hear the words. From the start every session is full of nominalisations by the client and therapist. In the sentence above there is: normally; own way; conscious part of you; front of your mind; distract itself; those stairs (no description of the stairs); more (doesn't specify how much more so the client has to figure that out)

D: And your unconscious mind that's normally at the back (separating conscious/unconscious with change in tonality, change in position my voice is coming from) can come to the front...and you can have an overwhelming sense of...***fully moving to the front***...with each step...that's it...and that unconscious part of you can...***increase in awareness***...of me of what I say...of the way that I say things...of tonality, subtle changes and can really...***fully become aware***...to the front of your mind...

Throughout the induction I am using the word 'and' to make sure that every part is linked. I am separating the conscious and unconscious as I talk to the client so that just by talking in a specific way and having my voice coming from a specific location the client will know whether I am talking to their conscious or unconscious mind or both. I am continuing to embed suggestions and commands to help the client's unconscious mind become more responsive to the way I am communicating to pick up on the marked out communication; including communication that is targeted at the unconscious mind like metaphors and stories.

D: Your conscious mind won't be completely to the back of your mind until you reach the tenth step...that's it...and the unconscious part of you won't completely be at the front of your mind...won't completely take over awareness until you reach that tenth step and you can be curious to discover what is at the bottom of those stairs,...that's right...that's right...(I lift up the right arm) and that arm can lower down only at the rate and speed that ***you go deeper***...and you won't ***go all the way down into a deep***

comfortable trance state until that arm goes all the way down...that's it...***all the way down***...that's it...that's it...(the arm has lowered now)

The main purpose of the steps is as a deepener and to create greater separation between what is coming up which is the laying down of the patterns for improved artistic ability and the initial conscious thinking. Normally when people use steps they count the client down the steps. This relies on the client going at the speed the therapist sets. By doing the arm levitation I have a signal that I can visually observe that is linked to the client going deeper at a speed they choose. I continue to use negative phrasing saying 'you won't ... until...' this generally creates less resistance and is slightly harder for the client; especially when they are relaxed; to unpick and analyse. Using negatives in this way people often hear the first part of the sentence (for example: You won't go all the way down into a deep and comfortable trance state) and if they are going to resist they often respond by doing it now and responding opposite to the statement. It is like a reverse double bind in that they can resist and respond now; or they can follow the suggestion and respond when they are asked to. The issue is about time not about whether they will do what they are being asked or not. It takes a lot for a client to then unpick the sentence and decide they will actually ignore it completely and not do anything. They are far more likely to choose - go against the therapist and do it now; or go with the therapist and do it when I am supposed to. Most people in therapy have made a commitment to be there to receive help so doing nothing when asked to is often not on their mind when they have an alternative to feel a sense of control and whatever other needs they would meet by going against the therapist.

D: And you can wonder where you are going to wander next...and somewhere there you can discover a painting...and I don't know what that painting will be of and what it is that it can teach you something about yourself...

Now that the client has reached the bottom of the steps I add a little confusion by using words that sound similar with one talking about thinking and one about action. I use many nominalisations as I don't know where they are going to be in their mind. And because I don't know where the painting that I want in their mind is going to be I don't tell them where it is; I give them the option to discover it. I keep using the word 'and' to continually compound end statement on the previous statement so that they build on each other. In many sessions I have observed people will say things like '...and in front of you, you will see a picture...' If the person can see in front of themselves in their minds eye before you have said that and there was no picture then you will mismatch their internal reality. By suggesting 'wonder and somewhere' it leaves the painting to be found. I also state a truism 'I don't know what that painting will be of...' This is undeniable with the implication that they must know. Being a truism they again will be in agreement with me. It is also followed with a poorly formed sentence. I run one sentence into the next using the words 'it can teach you' to transition from the end of a sentence with one meaning - saying ...what it is that it can teach you - then continuing the sentence giving a suggestion of what I want that painting to teach - ...it can teach you something about yourself. There is so much going on in the sentence with thinking about the wondering, finding a painting, hearing truisms that create agreement; that to also keep track of the final embedded command is difficult it just sinks in because analysing that as well takes considerable effort.

D: And there can be something curious about the painting...and as you **pay all your attention to me** so you can notice the things that I say and the things that I do and perhaps it will be the way that I say things...that's it...and I wonder what it is that's curious about the painting...

I am continuing with many nominalisations like 'curious' 'attention' 'things that I say' 'things that I do' 'way that I say things' 'wonder'. These help to keep the client on an inner search for their own meaning to what I am saying and doing. I sandwich the paying attention to me and emphasising to the client that there is something important in what I am saying and doing between two statements about being curious about the painting. This is partly to create amnesia for this paying attention. I want the client to unconsciously be responsive but to feel these bits haven't been said; that they were just listening to me talking about the painting.

Open Ended Suggestions

D: Could it be that movement and I wonder where that movement is and whether it is a little bit of movement or a lot of movement and whether that movement is in the centre or off to the sides or round the edges...and I wonder what type of movement it is whether it's a sort of wavy movement or a swirly movement or some other kind of movement and whether it has a 3d effect or a 2d effect...

While the client is paying attention to the painting I want to generate movement. I don't want to mismatch the client's inner reality (which could include movement or no movement in the painting). I decided to be fairly strong on saying that there is movement so I say there is movement but because I know the client could be seeing a still image I want to cover all bases with suggestions that lead to the movement being their but perhaps not at first observed. I mention the location and give the option for it being a little bit or a lot and the type of movement. I don't want to give a chance for the client to stop and think there is no movement so I say all of these options in fairly quick succession. I'm always implying there is definitely movement there even if at first it was so small it wasn't noticed until I mentioned it.

D: That's it and you can be curious to pay attention...to whether there are sounds there and perhaps they are coming from the picture or from elsewhere and maybe behind you or in front of you or to the left or the right or from above or maybe below...that's it...that's it...that's it...

This is a technique I use quite often; where I will suggest an idea as if any possibility is an option then follow a preferred route. For example above I suggest paying attention to see if there are sounds there; giving the choice that there may be silence but before the client has time to think about it I suggest that sounds are there it is now a question of where they are coming from. I then go on to give multiple options of where the sounds could be originating.

D: And another curiosity about this picture is that you can step inside this picture...that's it...and I don't know what it's like the other side of the picture...

When I said 'step inside the picture' I could see the client had done that by changes in his physiology which I acknowledged by saying 'that's it' I then followed this up by immediately stating a truism that I don't know what it is like the other side of the picture again implying that they do. And for the client to know what it is like they have to be there. By having the client step into the painting it takes them even deeper into trance. Any time you have someone change where they are in their mind they go deeper and layer their trance experience so that the last place they go gets sandwiched between the previous places, which are sandwiched between the places before that etc... This leads to usually spontaneously getting amnesia for the deeper parts of the trance.

Metaphors for Unconscious/Conscious Processes

D: That's it...you know normally...that conscious part of you...guides your decisions...guides what you do, what you're thinking about...it's a bit like a driver of a train...the driver is just a small part of the whole thing...and the driver can just see what is outside his window...and the driver knows that there are 8 carriages behind and the driver knows that each of those carriages is full of people and that each of those people are saying their own things and doing their own things and each of those people are in control of what they're in control of...some are reading newspapers...some are listening to music and some are planning ideas and many of them are lost in thought...that's it...yet all the driver is aware of even though the driver knows all of that is there...is what's through the window...

I have used a metaphor to parallel the conscious mind by having the conscious mind as the driver of a train and the passengers as neurons that are all doing their own thing independent of the driver. People consciously know their mind is doing lots of things at once that they have no awareness of; and that all they are aware of is what they consciously are currently aware of. Like the driver only being able to see out the window. I am conveying this message as a metaphor as this is an easier way to lay down a pattern for the unconscious mind to use.

D: That's it...whereas the unconscious part of you is like a super being floating above the train like superman flying above the train where he can see the passengers...because he can see through the walls because he can see what they are all doing, he can notice their behaviours their language...he can fly down and talk to them...he can even make them change their behaviours...he can ask someone to stop reading their newspaper and they would stop...he could ask someone to stop listening to music and they would stop...he could interrupt someone having a conversation and they'd forget what it was that they were talking about...

To describe the unconscious mind I use the term super being as it has many positive connotations; I describe some of the strength of the unconscious mind and what it can influence by talking about how it can influence the passengers. The idea I want to convey is that a conscious knowing and an unconscious doing are two different things. I want the client to be able to 'just do' on an unconscious level.

Post Hypnotic Suggestion

D: That's it...**now you can be curious**...as to how you are going to **use all of your unconscious resources**...how you are going to use all of the talents you've got that are normally held back behind the doors of the conscious part...you can be curious about what the improvements will be...and how your mind and your circuitry of your brain will make aspects of the improvements permanent in a comfortable way...that's it...that's it......and you can get a sense of what it is like to see you in your mind...to see you in the future...

There is implication running through this section. When you use implication or presuppositions they act like post hypnotic suggestions. By saying 'you are going to' and 'will' it is placing what I am saying in the future as a certainty rather than a possibility. I follow the suggestions up by building on each preceding suggestion; so I firstly state what will happen in the future, then I move on to 'see you in your mind' then onto a context for that 'you' that is being seen - in the future. If I didn't add that the client could see themselves at any age and any time even made up ages and times (like in the distant past or a futuristic world). I want to keep what I say as unthreatening as possible so I build one thing on another hoping to go just slightly faster than the client's awareness of what I am suggesting so that I am leading their internal reality now. In the same way that you can wear glasses and not notice until someone mentions them; I want the client to assume they've just not noticed something until I mentioned it; rather than it not previously being a part of their reality.

D: And you remember Superted...Superted used to say his magic word and he would change from an ordinary teddy bear into a super-teddy bear...and consciously you've watched Superted and you could never quite hear the word…you'd never know consciously what the word is no matter how often you watched it…and you can see yourself…and you can see that you having a code word…and you can watch yourself say that code word to yourself…you can watch yourself say that code word to yourself…

I now use the cartoon character Superted to introduce the idea of a code word for triggering the artistic ability. Ideally I want the code word to be something internal and unconscious rather than something the client has to consciously say. It is like when a hypnotist sets up the word 'sleep' to re-induce trance. I want the client to have a word to re-induce artistic abilities but I want it to come from the client's unconscious mind. To cement ideas and reprogramming I normally give the idea; then have them watch themselves doing the new behaviour; then have them go into that version of themselves to experience actually doing what they have just watched themselves do. This is generally a very effective way of getting that behaviour into the client's future. It also matches the way people normally do things. They get the information (me giving the idea); then think about what they are going to do (see themselves doing the behaviour); then they do it (stepping into that future them and experiencing it). If you just jump someone into just doing a new behaviour it may not stick because there was no planning or mirroring reality. Also by doing it this way if there are any problems with the way they see things go it can be changed to be just right before they go into the experience.

D: And when you watch yourself becoming overwhelmed by that compulsive artistic ability...and I don't know if it's thousands of times or even tens of thousands of times or even realistic and lifelike...in comparison to how the conscious part of you...that other part of you does art...that's it...that's it...and you can watch and I wonder what you see...and I wonder what you **see how well you're doing** that art...does it look haphazard to start with like when you watch Rolfe Harris...where that other part...the unconscious part has its own way of drawing where **it takes control...it takes control**...and I wonder whether that you reports that it's like the hand doing all of that work themselves...whether it's like an image just being printed onto a page straight from the mind...whether it's like **the hands just get a compulsive feeling**...whether it's like **the hands get a compulsive feeling to carry out that artistic talent**...

People with 'natural' artistic ability or people that are 'naturally' talented at anything generally have a high level of compulsion to do what they are interested in. In this section it is this compulsion that I am working at installing. I offer lots of ideas about how that compulsiveness will take effect and that it comes from the unconscious part of the client not the conscious part. I do this by taking about the hands doing the work and that the unconscious mind has its own drawing style. The presupposition through all of this is still one of that improved drawing ability taking place; it's now a question of what it will be like to the conscious mind as an observer rather than whether it will happen. I am also talking to the client as someone that is still observing that future them. They are not yet in that future them.

D: And I wonder how long that lasts...is it for 20 minutes or half an hour or is it for a full hour...and I wonder whether it ends because you decide it's time to stop or whether it ends because the time is up...and I wonder how long it lasts...is it twenty minutes, half an hour or maybe even a full hour...or somewhere in-between...

I now limit the duration of that compulsive behaviour. I don't want it to carry on indefinitely because that could have negative effects on work and family life. I want the client's unconscious mind to decide the duration so I offer choices around how long it will last rather than just telling the client how long it will last. Whenever doing any work with clients the therapist needs to be mindful of the positive and negative effects of making the change. By having the client view the changes first it gives them the opportunity to see if the changes are acceptable. If they are not you have given the client the option to decide not to go into themselves when we get to that stage and also to make any changes now before they do. It is a bit like starting with hindsight; the client can see if the changes are acceptable from a dissociated position. Being dissociated gives a useful view on things. Like watching a football match and knowing a player should have passed because you saw another player open; yet the player that should have passed wasn't in a position to see what you could see so made different decisions.

D: And like all resources...you can use...the skills and talents and you can transfer to other areas...and that's **TRANCEfer to other**

areas...that's it...excellent...and you know what it's like to...***step inside that you there and experience that deep compulsive desire to create photorealistic art work***...and I wonder what changes happen at a neurological level that are completely comfortable and healthy that build the talent and increase the talent by temporarily numbing down an area of the brain and I wonder how that numbing down takes effect...

I use the term TRANCEfer to imply going into trance to transfer skills and abilities.

Shutting Down Perceptual Filters

D: You can be curious as to whether it will be like certain signals not being allowed through or certain signals taking a different route around the brain...performing those talents with those signals taking different route to how you perform it consciously and you can imagine what it is like to be that you there carrying out...that's it...that ***artistic compulsion***... and I wonder what it feels like...do the hands feel tingly...or is it in the arms or is it on the back of the neck...

I want to generate feelings associated with the specific trance of improved artistic ability so I ask what it feels like then I tell the client in the next sentence what it will feel like; that it will be a tingling. I them make the focus on where that tingling will be rather than if there will be a tingling or not. I am also setting up the next part of the session where I want the tingling to be link to energy.

D: You can be curious as to how **that compulsion takes effect** once you hear that word and I don't know whether the word is Leonardo, or whether the word is Rafael or some other famous artist...that's it...that's it......that's it...and you can **go deeper...and deeper**...and you know what somnambulism is......that's right...and when that **artistic desire...compulsion**...takes effect I wonder whether it makes the fingers twitch with nervous energy desperate to carry it out or whether that doesn't show...

The self-hypnotic suggestion is bought back in again here and linked to the feeling/energy. I now make the focus on whether the energy will be visible to others or not rather than if it will be there or not. When I introduce something I want to be there I like to make the focus on something to do with that new thing rather than if that new thing will be there or not.

D: And you can imagine in your mind a panel and it's a panel with many levers on...and each lever is logically labelled...that's it (he moved)...and you can **imagine turning up that lever to artistic ability**...putting it up to full...that's it...that's it...and in a moment I'm going to lift up your right hand and when I do I'm not going to tell you to put it down any faster than you...**become absorbed**...in the idea of being a great artist...and I don't know whether it's going to be an artist that is going to be a mixture of many artists or an artist that is going to be greater than any artist you've ever known...that's it...and I don't know whether your hand coming down will take two minutes, three minutes or five minutes or somewhere in between...and I don't know how long that will be on the inside...for you...and I wonder what you get up to while that is happening......

The client already has the experience of using arm movement to go deeper into a trance which is becoming more absorbed. I now use it again to become absorbed in the idea of becoming a great artist. Again I introduce an idea; the idea of becoming deeper absorbed; then I focus attention on time rather than whether the absorption will happen or not and then change focus to what the conscious mind will be getting up to.

D: And when it's time to come out of a trance and come all the way back to the room you can **come back with the artistic abilities**...and I'll let you know when that time is...and I wonder what it will have felt like to have those **changes occurred on a deep neurological level**...first changes can occur though all the connections, pathways, circuitry...that's it...and on one level when you come back how you can be curious as to where those changes came from and how those changes occurred also...

The suggestions given here are given very directly that when the client comes out of the trance they will come back with the artistic abilities. I use the word occurred rather than occur in the middle of a sentence to imply they have happened; then I mention a curiosity about where those changes have come from to again change the focus from whether it happens or not to trying work out where the changes came from.

D: And I wonder whether you will be amazed or at least shocked...and you know the studies that have been done by **shutting down areas temporarily in a healthy way**...whether it is shutting down or just not allowing the signals to get through temporarily from the logical rational hemisphere...part of the brain...the creative part of the brain the part that notices fine detail notices every little thing, notices millions and millions of bits of information every single second...**that part of the brain can take control**...and then it will end up with you having...(client moved his arm to scratch)...that's it...many abilities...that's it...that's it...now as you're going to achieve great things...that's it...that's it...we want you to take your time to do this...**take your time to do this now**...that's it...that's it...(I lift the clients arm above his head)...that's it...that's it...and on many levels you can begin to count backwards from 100 and **let the unconscious work**...that's it...and you know what your right foot feels like and I wonder if it feels different from the left...that's it...that's it...and your left hand can be left there lightly resting where it is and I wonder how your eyelids feel...that's it...that's it...making this completely...

While the client has their arm in the air and is counting down in his mind I want to jump his attention all over the place to stop him focusing on the lowering arm and the internal unconscious work.

D: That's it and you know a minute of my time can seem like a longer time of yours just like you can experience...that's it...(arm fully came down)...a longer time that just goes by from minute to minute...and you can take a minute of my time to go deep and comfortable inside your mind into a deep and comfortable focused state of mind...***allowing that artistic ability to develop and enhance itself*** and the more it enhances itself the more pleasure you can experience inside your mind and you can take a minute of total silence to do that now...(minutes silence - I just sit observing minimal cues)...

I sit and observe and watch for signs of increasing pleasure to see that he is enhancing his ability.

D: That's it...that's it...and you know you can control blood flow, you can make yourself blush on half of your face and if you get a cut you can make it so that the blood stops flowing over the cut any more than is necessary to keep the wound clean...you can make an arm numb or a leg numb or half a head numb...you can increase your metabolism or slow it down...you can alter any system in your body with your mind...

To link back with the earlier statement of making part of the brain numb I sandwich the suggestion for this again in the middle of a collection of truisms about what the mind is capable of.

D: Now when it's time for you to open your eyes I'll ask you again to draw a horse and again you will get one minute to do it in...and you can be curious as to how much better that horse will be drawn...will it be lifelike, will it be hundreds or thousands of times improved on the last one...how much can you manage to draw in one minute...you can be curious as to how much you can draw of that horse in one minute and the level of detail you can draw...and **you can always keep in mind that ability** those abilities you have...and **you can always keep in mind how to get that artistic ability**...and **you can always keep that in mind**......that's it...and you can open your eyes now...

My use of the term 'you can always keep in mind' is specifically used because if you always keep something in mind it means it is always there. It is a post hypnotic suggestion to make everything stick.

D: (Now talking very normally) Right want to have a bash at drawing the horse again...we'll see how it goes now...help if the pen works (pen not working so the client changed pen)...(one minute given to re-draw the horse)...times up...how do you think you did?

C: It looks more like a horse than the last one did; it's more proportionate I think

D: Yes, drawing style was different as well

(client starts to carry on drawing)

C: I'm not supposed to be doing this now am I?

D: No, I'll show this to the camera before we...if you do start finishing it off or something

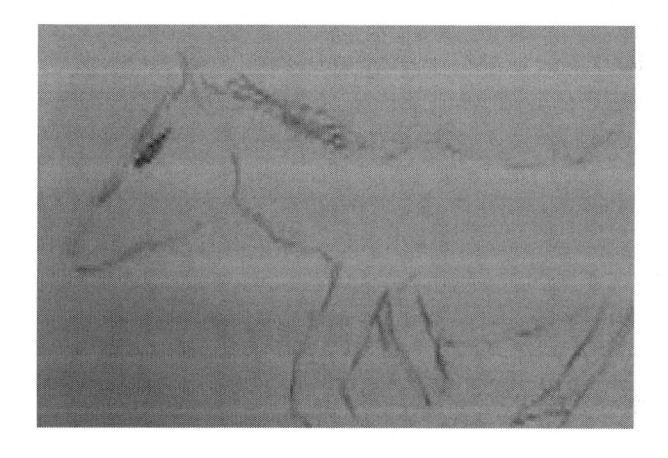

C: That is quite different isn't it (client picks up a pen again to carry on drawing)

D: And that's only a first thing drawn after being zapped

Next few minutes was spent me just watching as the client compulsively kept drawing, putting the pen down to talk then carrying on getting absorbed in drawing again, getting more involved in what he was doing and adding more and more detail, then adding colour and motion. In total he spent about 4 minutes on the picture adding to it, putting the pen down for a few moments to talk but then not talking instead he would pick the pen up and carry on drawing. I tried to get his attention to discuss his experience but he struggled to tell me as he was more absorbed in continuing to draw. Below is the picture after about four minutes of drawing on it.

After the session the client went home and drew an image off of a Disney video case. He said it only took him a few minutes to sketch it. He then put it on his computer and coloured it in. He said he was amazed at what he had done and didn't understand how it was possible.

211

Printed in Great Britain
by Amazon.co.uk, Ltd.,
Marston Gate.